Integrating Design and Manufacturing
For Competitive Advantage

INTEGRATING DESIGN AND MANUFACTURING FOR COMPETITIVE ADVANTAGE

Edited by

Gerald I. Susman

New York   Oxford
OXFORD UNIVERSITY PRESS
1992

Oxford University Press

658.5752
I61

Oxford   New York   Toronto
Delhi   Bombay   Calcutta   Madras   Karachi
Kuala Lumpur   Singapore   Hong Kong   Tokyo
Nairobi   Dar es Salaam   Cape Town
Melbourne   Auckland

and associated companies in
Berlin   Ibadan

Copyright © 1992 by Oxford University Press, Inc.

Published by Oxford University Press, Inc.,
200 Madison Avenue, New York, New York 10016.

Oxford is a registered trademark of Oxford University Press.

Library of Congress Cataloging-in-Publication Data
Integrating design and manufacturing for competitive advantage /
edited by Gerald I. Susman.
p. cm.   Includes bibliographical references.
ISBN 0-19-506333-3
1. Design, Industrial.   2. Engineering design.
3. New products.
I. Susman, Gerald I.
TS171.4.I575   1992
658.5'752—dc20   91-36271   r 91

TP

9 8 7 6 5 4 3 2 1

Printed in the United States of America
on acid-free paper.

PREFACE

This book is about managing the new product development process, particularly about integrating the design, development, and manufacturing functions so that high quality products can be introduced to the market faster and at competitive prices. Customers, suppliers, and internal company functions, such as marketing, purchasing, and accounting, also need to be integrated into the new product development process, but the dynamics of inclusion and integration of all relevant players can be understood by focusing on the design-manufacturing interface. This interface is, perhaps, the most complex of those encountered in cross-functional product development teams. Mastering this interface is more than adequate preparation for dealing with the others.

The study of the design-manufacturing interface begins with appreciation of the strategic capabilities that companies need to develop in order to manage this interface effectively. Several chapters of this book discuss strategic capabilities, such as the systematic development and introduction of new technologies into products and processes and the use of appropriate tools and techniques to facilitate communication and problem-solving between design and manufacturing personnel. Also discussed is learning from experience, a strategic capability that companies can strengthen and accelerate by providing sufficient opportunities for such learning to occur. These strategic capabilities provide a basis for understanding the specific actions that key product development personnel take and increasing the likelihood that their actions will be successful. Several chapters also discuss the social, political, and cultural context within which these key players interact. This context can facilitate or inhibit the prospects for successsful integration between functions.

The study of the design-manufacturing interface also requires appreciation of the degree to which the new product development process is information intensive. Managing the type of information to be processed and developing an effective structure for processing information are critical to new product development success. The information intensive nature of the product development process is demonstrated throughout the book and prompted the development of a model in the book's final chapter that links the role of information in the product development process and a company's capability to organize, process, and learn from that information.

Competition in many domestic and international markets appears to be entering a new phase, in which product quality and performance are becoming more important to customers than price. Companies that can introduce superior products to such markets sooner will earn higher profits and gain market share at the expense of their slower competitors. In such markets, the effective management of the new product development process is the essence of competitive advantage.

University Park, Pennsylvania
December 1991

G. I. S.

ACKNOWLEDGMENTS

This book was planned as a counterpart to the Klein Symposium on the Management of Technology that was held at Penn State in August 1990. The book was intended as an opportunity for researchers to present their current ideas and research on ways to improve the new product development process. The Klein Symposium was designed as an opportunity for researchers and executives to share ideas and experiences on the same subject. Although the book and symposium were separate projects, I had hoped for maximum synergy between them. I believe that this hope has been realized.

The contributing authors mailed copies of their chapters to each other and came to Penn State two days prior to the start of the symposium. They discussed and critiqued all the chapters, then stayed to hear presentations by eight senior executives and to participate in discussions with them. In turn, the executives received abstracts of the authors' chapters prior to their arrival at Penn State and commented on them during the symposium. The executives influenced the book chapters in both subtle and explicit ways. When the latter was the case, the authors cite the appropriate presentations in their chapters. However, I thank all the executives for their general influence on the book. They are Satish Agrawal, Polaroid; Ronald B. Campbell, Jr., Xerox; James Coraza, IBM; Larry Dittmann, AMP; William E. Hoglund, General Motors; Richard Sphon, Corning; Louis C. Varljen, Armstrong World Industries; Joseph W. Yanus, Alcoa.

I thank Robert and Judith Klein, whose generosity made the Klein Symposium on the Management of Technology possible. They contributed immeasurably to the quality of the book by providing me with the funds to invite all the contributing authors to work on their chapters face-to-face and to participate in the symposium. Also, their warm personalities have made the transition of their role as benefactors to that of good friends seem natural and effortless.

I also thank Judith Sartore, Supervisor of the Research Publications Center at the Smeal College of Business Administration at Penn State. Most professors would be pleased simply to receive a typed manuscript on time with few typographical errors. It is exceptional, however, to work with someone who has excellent editing skills and the capability to attend to details that would be overlooked by almost anyone else. Barbara Apaliski's able assistance in preparation of the manuscript deserves recognition also.

Finally, I thank my wife, Liz, but not for the reason often given in acknowledgments, that is, for patience and understanding while I worked on this book. She also is a very busy academic and I have to muster as much patience and understanding for her projects as she does for mine. Instead, I simply thank her for being who she is and for bringing out the best in me.

CONTENTS

PART III   SOCIAL, POLITICAL, AND CULTURAL CONTEXT

CONTRIBUTORS

Paul S. Adler
University of Southern California

Philip Barkan
Stanford University

W. Bruce Chew
Harvard Business School

Kim B. Clark
Harvard Business School

Paul D. Coughlan
London Business School

James W. Dean, Jr.
University of Cincinnati

Arnoud De Meyer
INSEAD

Ronald J. Ebert
University of Missouri, Columbia

John E. Ettlie
University of Michigan

Mitchell Fleischer
Industrial Technology Institute

Arthur Francis
Imperial College of Science
Technology and Medicine

Takahiro Fujimoto
Tokyo University

Jeffrey K. Liker
University of Michigan

Stephen R. Rosenthal
Boston University

Susan Walsh Sanderson
Rensselaer Polytechnic Institute

Gordon V. Shirley
University of California, Los Angeles

E. Allen Slusher
University of Missouri, Columbia

Gerald I. Susman
The Pennsylvania State University

Mohan V. Tatikonda
Boston University

Diana Winstanley
Imperial College of Science
Technology and Medicine

Integrating Design and Manufacturing
For Competitive Advantage

INTEGRATING DESIGN AND MANUFACTURING FOR COMPETITIVE ADVANTAGE

GERALD I. SUSMAN

This book focuses on managing the design, development, and introduction of new products, with particular emphasis on how design and manufacturing can work together to manage these processes effectively. There are at least two reasons for this focus. First, competitive pressures are leading companies to offer their customers higher-quality products at lower prices and to deliver these products more quickly. These goals cannot be achieved simultaneously only by improving efficiencies within each organizational function. Simultaneous advances in these goals will require systemic changes in the way these functions relate to each other. Second, technological breakthroughs are increasingly becoming a significant driver of competitive advantage in new as well as mature industries. Such breakthroughs can lead to improved product performance. A major issue concerns how to introduce new technology into new products in a disciplined and strategically focused manner.

It traditionally has been assumed that low price, high quality, and short lead time cannot be achieved simultaneously; that is, one goal has to be traded off against another. Recent innovations in manufacturing practices, however, indicate that the manufacturing process can be managed so that a trade-off between these goals may not be necessary. For example, total quality management programs have demonstrated that cost and quality can be achieved simultaneously (Crosby, 1979); just-in-time management has demonstrated the same for cost and variety (Schonberger, 1982, 1986); and, recently, Stalk and Hout (1990) have demonstrated that cost and lead time need not be negatively related. As De Meyer suggests in this volume, the issue is not whether such a trade-off exists, but rather that most companies are far from the efficient frontier where such trade-offs become necessary.

Effective management of product and process design offers yet another potential source of simultaneous improvement in cost, quality, lead time, and performance. These improvements result from a better fit between product and process design and from greater simultaneity in the design of the product and process. Both of these efforts require a significant change in the relationship between the design and manufacturing functions. Also, the type of technology that is introduced into new products and when and how it is introduced affect product performance and the ability of design and manufacturing to work together during the product development cycle.

This book is devoted primarily to exploring the issues associated with effective management of the product development cycle and to the nature of the organizational and technological changes required to facilitate and maintain that effectiveness. The issues to be explored focus primarily on design for manufacturing. Design for manufacturing is explained briefly in this chapter, as are strategies for competitive advantage that can benefit from its effective implementation. The remainder of this chapter and the remaining chapters of this book are organized around three organizational issues that influence or are influenced by effective design for manufacturing. These issues concern an organization's strategic capabilities, the role of design for manufacturing in the new product development process, and the social, political, and cultural context within which design for manufacturing operates.

Twenty experts on the management of technology have contributed fifteen chapters to this book. These experts are Paul Adler, Philip Barkan, Bruce Chew, Kim Clark, Paul Coughlan, James Dean, Arnoud De Meyer, Ronald Ebert, John Ettlie, Mitchell Fleischer, Arthur Francis, Takahiro Fujimoto, Jeffrey Liker, Stephen Rosenthal, Susan Sanderson, Gordon Shirley, Allen Slusher, Gerald Susman, Mohan Tatikonda, and Diana Winstanley. The chapters of these contributing authors focus primarily on one of the three organizational issues discussed above. However, they may have addressed more than one of them, as they were not required to write exclusively on a single issue.

DESIGN FOR MANUFACTURING[1]

The term *design for manufacturing,* or DFM, is used in this book to characterize efforts by design and manufacturing to improve the product-process fit or to increase the degree to which the product and process are designed simultaneously. These efforts may take the form of changes in organizational practices. For example, manufacturing may be involved earlier in the product development cycle than has been the case traditionally. Manufacturing personnel may suggest how to design the product for ease of manufacturing. It is much less costly to incorporate their suggestions into the product in the early phases of the product development cycle than after the product design has been released to manufacturing. A more ambitious step is for design and manufacturing to develop the product and process simultaneously. Manufacturing may gain enough knowledge about the product design during early involvement to proceed with the process design. This latter effort requires both functions to coordinate their activities closely during the overlapped phases.

Manufacturing personnel can make their contributions to the product development cycle most effectively by use of cross-functional coordination mechanisms and/or by use of specific tools and practices. The coordination mechanisms offer opportunities for manufacturing personnel to communicate their inputs to design personnel either face-to-face or through a designated intermediary. Some tools and practices can supplement or even substitute for communication between design and manufacturing personnel on design for manufacturing issues. Other tools and practices can be used early in the product development process to select appropriate product and process technologies or to clarify design concepts for a single product or for a family of products. Rosenthal and Tatikonda, Sanderson, Barkan and De Meyer discuss these tools and practices and their strategic potential in detail. They and other authors also discuss the prerequisites to their effective use.

COMPETITIVE STRATEGY

The value of design for manufacturing can be appreciated better if viewed within the context of competitive strategies that can benefit from its use. A competitive strategy can be used to articulate the mission of a research and development program, offer guidance for a clear product definition at the start of a new product development project, and provide a rationale for the selection of specific DFM tools and practices.

Susman and Dean (1989) offer three strategies for competitive advantage that make particularly effective use of design for manufacturing in an environment that increasingly restricts the leeway of firms to trade off cost, quality, lead time, and product customization. This environment is also increasingly turbulent owing to increased industry segmentation and shorter product life cycles. A prerequisite to the effectiveness of DFM tools and practices in all three strategies is the simplification of the production process, elimination of non-value-added activities, and improvement of product quality.

The first strategy is competing on multiple dimensions, i.e., low cost, high quality, and short lead time, so that companies can differentiate their products on more than one competitive dimension. This strategy is particularly useful for companies in mature markets in which commoditylike products typically compete on the basis of price. DFM can contribute to lower cost, higher quality, and shorter lead time by involving manufacturing earlier in the new product development process.

The second strategy is competing in multiple segments. In this strategy, companies differentiate their products by market segment and seek to earn a premium price in each segment by tailoring the product to the segment's customers. DFM can contribute to lowering the costs of variety by designing product families with a high percentage of common parts or modular parts. Sanderson discusses the development of a family of Sony Walkmans; each product in the family was designed for consumers with different lifestyles and varying ages. The development of product families in which each product serves different market niches is an example of development of a family of products in a horizontal direction.

The third strategy is competing by continuous product improvement. In this strategy, the companies frequently introduce new and improved product models and earn a premium price for them because their products are superior to those of their competitors and difficult for them to copy and sell at a lower price. Sanderson discusses how five core technologies can be introduced incrementally into successive generations of the Sony Walkman. Sony's plan was based on a vision that the Walkman would have the highest-quality sound in the smallest possible package. The development of product families in which each successive generation is superior in performance to its predecessor is an example of development of a family of products in a vertical direction.

STRATEGIC CAPABILITIES

Several of the chapters discuss strategic capabilities that companies can use to support the strategy they have chosen to pursue. A strategic capability offers a company a sustained competitive advantage when substantial time and effort is required for competitors to develop the same capability. A number of such strategic capabilities can be identified in the chapters of this book.

Technology Management

While the United States still leads the world in research and development, Japan and Germany have exploited the fruits of research and development much more effectively by their superior capability in managing the introduction of technology into new products (Gomory, 1989). This capability includes developing a vision for a product concept and diligently applying organizational resources toward realizing the vision, benchmarking the competition's product development capabilities and selecting core technologies that will meet or exceed those capabilities, assessing the readiness of new technology for introduction into products, assessing the remaining life cycle of existing technologies, and developing product families that include technological advances in each new product generation.

Barkan contrasts the demands and skills required for managing research and development with those required for managing product engineering. These two functions serve very different organizational objectives. R&D serves long-run objectives. Breakthroughs are infrequent and a high failure rate is expected for development of new product and process technologies. Product engineering serves relatively short-run objectives; it should be approached incrementally and a low failure rate should be expected. Progress on both objectives should proceed simultaneously but in a complementary fashion. R&D should concentrate on developing and experimenting with new technologies until they are proven. Product engineering should introduce new technology incrementally into new products only after they have been proven or at least are "ready" for such introduction. Companies can get into trouble by thinking that most new products should hit "home-runs" when they are introduced rather than hit a series of "singles" over successive generations of products.

The possession of proven core technologies can be a strategic capability if research and development and product engineering are managed to complement one another. Previous mention was made of Sanderson's case study of Sony in which several core technologies were developed over a number of years and introduced systematically into new products. New technologies were introduced at frequent intervals because the life cycle of each product generation was relatively short. Barkan cites examples of Japanese companies that reduced project risk and assured a smooth product development cycle by systematically introducing proven core technologies into new products. Barkan also reviews some of the tools and techniques that can be used to assess whether new technology is proven and ready for introduction into new products.

Tools and Practices

Several authors have pointed out that management of the product development process is information intensive. As such, companies with superior information-processing capabilities can enjoy a sustained competitive advantage by using these capabilities in support of strategies that exploit them. The three strategies cited above are ideally suited for use of these capabilities. Rosenthal and Tatikonda identify six information-processing functions that can enhance an organization's ability to implement these strategies. These functions facilitate cross-functional integration and increase the efficiency and effectiveness of the new product development process. Rosenthal and Tatikonda introduce a number of DFM tools and practices that can help organizations to develop the six information-processing functions.

Sanderson introduces a number of design tools and information-processing capabilities that can help an organization manage the continuous or evolutionary design of products over the life cycle of a family of products. Barkan refers to several structured methodologies and disciplined procedures that can help an organization manage the pattern of engineering-change notices so that they occur earlier in the product development cycle. Finally, De Meyer discusses the extent to which European companies currently use or are planning to use various tools and practices to improve the speed with which new products are developed.

Shirley introduces the concept of modularity and its relationship to competitive strategy. Modularity is an approach to design problems that attempts to decompose them into sub-problems that have minimal interaction. Products and processes that are designed with modularity in mind have sub-products and sub-processes that are relatively self-contained. DFM is considerably simplified for products and processes that are amenable to modularity because DFM adds considerable constraints to the problem of designing these products and processes. Modularity simplifies DFM by reducing the degree of interdependence between designers and between them and manufacturing personnel. Several authors in this book use interdependence as a key attribute in characterizing the design situation that design and manufacturing personnel face. The more the products and processes these personnel design can be modularized, the easier their common task will be. Shirley also relates modularity to minimizing the product development costs associated with pursuit of the three competitive strategies identified earlier and to facilitating learning across horizontal and vertical product families.

Organizational Learning

The ability and speed with which a company can learn from experience is another strategic capability. The ability to learn is dependent, in part, on how the company captures and accesses information. Companies can simplify this process by minimizing the amount and complexity of information they have to process. Clark, Chew, and Fujimoto as well as Adler and Barkan cite examples of companies that set limits on the number of new technologies that can be introduced into any new product or minimize the number of new product parts that can be introduced between adjacent product generations.

Barkan cites several reasons that organizations fail to learn from experience, such as an unwillingness to revisit past difficulties that in hindsight appear avoidable, to preserve and share information, or to challenge the authority of those responsible for past decisions. Clark, Chew, and Fujimoto cite the speed of feedback from multiple design-build-test cycles, especially during the production of prototypes and dies, as critical to organizational learning.

Sanderson discusses the use of multicycle concurrent engineering and virtual design as methods by which knowledge of existing products and processes can be used to project the future development of products within existing product families. Finally, Coughlan describes how the pattern of rotation among project personnel within a company can facilitate or impede learning opportunities across successive generations of products within a family.

Supplier Network

No strategic capability is more important or time-consuming for a company to develop than a strong supplier network. Early involvement of suppliers in the design of products and processes can reduce total product development time. A trusting relationship between a company and its suppliers can reduce bureaucratic paperwork between them and increase their willingness to share ideas. Clark, Chew, and Fujimoto demonstrate how essential a strong supplier network is to the rapid production of prototypes and dies, which, in turn, contributes significantly to reducing total product development time. De Meyer documents the extent to which European companies are expanding their supplier networks in an attempt to improve the speed of new product development. Liker and Fleischer discuss some of the social, political, and cultural impediments to effective development of such networks among United States companies.

DESIGN FOR MANUFACTURING AND THE PRODUCT DEVELOPMENT PROCESS

DFM can contribute to the product development process in different ways and at different times, depending on decisions made before or shortly after the product development process begins. These decisions include the choice of technology to be used in the product; whether the product is the first member of a family of products or one-of-a-kind; and the level of difficulty of performance, cost, quality, and schedule goals. These decisions influence the amount of uncertainty and complexity that manufacturing and designer personnel face when the product development process is initiated as well as the degree of interdependence they must manage. These outcomes, in turn, influence the nature of the inputs that these personnel can make and when they can make them most effectively.

Ettlie discusses the critical role of the concept development phase in the product development process. Uncertainty is at its peak when this phase begins, but so is the potential for creativity. New products are conceived and more mature products are revitalized during this phase. Customers are identified for products and product attributes and prices are set to meet customer expectations. A clear product definition is critical to the success of all remaining product development phases. Ettlie explores the respective roles played by function representatives during the concept development phase and their relationship to project success. He also presents data that suggest that the greater the percentage of total project time that is devoted to concept development, the greater the overall project success. The ratio of manufacturing engineers to design engineers in the company appears to influence the percentage of time devoted to concept development. Also, the greater the degree of design-manufacturing integration that exists within a company, the more time that project personnel devote to concept development.

Slusher and Ebert focus on the product design phase and identify four types of design situations that product designers might face. These situations vary by the degree to which the design problem is structured and by the frequency of problems that can occur during the product design phase. Slusher and Ebert's typology can be used at the start of the new product development process to determine the type of organizational structure and managerial style that is most appropriate for dealing with a particular type of design situation. Their typology also can be used to assess changes in the design situation over time as a consequence of technology management decisions or other decisions made before or during the product development process. An organization's ability to learn can spare designers from having to face repeatedly the same type of unstructured or high-problem-frequency design situation.

Adler develops a typology that is similar to that of Slusher and Ebert, but he uses it to determine the mechanism that is most appropriate for coordinating the design-manufacturing interface and when it is most appropriate to use the mechanism. The conditions that determine these choices are the novelty and the analyzability of the product-process fit. Novelty concerns the number of fit problems that need to be resolved. The greater the degree of novelty, the more the coordination mechanism should rely on interpersonal interaction, i.e., use of teams rather than standards or plans. Analyzability concerns how readily available a solution is to the fit problem. The more difficult the analyzability, the greater the coordination burden will be on later phases of the product development cycle, i.e., at the production phase rather than earlier.

Coughlan discusses how the deployment of manufacturing engineering resources can influence the number of manufacturability-related changes that occur late in the product development process, i.e., after the start of volume production. His study suggests that manufacturability-related changes are lower when experienced manufacturing engineers (MEs) are assigned to first-of-family products. He cautions, however, that if experienced MEs were assigned only to first-of-family products, learning opportunities might be lost by failing to assign the same MEs to successive generations of the same product family. He also presents data that suggest that MEs can be deployed too early in the product development process when the product is not the first-of-family. The MEs may freeze the design prematurely in such cases. The company that Coughlan studied prioritized its efforts to ''design right-first-time'' for attributes that were evident to customers, but ''design right-next-time'' for manufacturality-related changes that were evident to manufacturing, but not to the customer.

Clark, Chew, and Fujimoto suggest that a significant contributor to the Japanese advantage in new product development lead time and engineering productivity is superior manufacturing capability. They point out that a significant portion of the product development process involves the manufacture of prototypes and dies as well as pilot production and ramp-up to volume production. The speed and efficiency with which Japanese automobile manufacturers are able to build prototypes and dies increases the number of design-build-test cycles that can be conducted on the product and permits a longer process development phase in which to refine the product-process fit. As a result, Japanese manufacturers spend much less time on pilot production and ramp-up than do manufacturers in the United States and Europe. These authors also identify similarities between design and manufacturing activities and suggest that improvements in the activities of one function are generalizable to those in the other.

SOCIAL, POLITICAL, AND CULTURAL CONTEXT OF DESIGN FOR MANUFACTURING

A dominant theme in many of the chapters discussed thus far is integration of organization functions and integration of the organization with key elements of its environment, e.g., suppliers and vendors. Integration remains a theme in the chapters in this section, but primary attention is paid to differentiation. Kingdon (1973) reminds us that differentiation can occur not only between lateral functions, but also between hierarchical levels and between program management and functions. The social, political, and cultural context within which American and European companies currently operate tends to foster differentiation between internal organizational units and levels as well as between the organization and its environment. The introduction of DFM into an existing organization demands significant innovation to manage or reduce differentiation, but such innovation is unlikely to occur unless the social, political, and cultural context has been modified to promote it.

Susman and Dean focus primarily on facilitators of effective DFM. They developed a model based on interviews with personnel from twenty-one projects in eleven companies. These interviews identified practices these companies undertook at the organizational and project levels to facilitate DFM. The practices were clustered into three categories: integrative mechanisms that can be implemented organization-wide; group processes that facilitate group problem-solving; and codification and computerization of data on manufacturing requirements and design guidelines. Integrative mechanisms serve mainly to reduce differentiation, while group processes and codification and computerization of data are alternative means for processing information related to DFM. The model developed by Susman and Dean hypothesizes relationships between categories of practices and overall project success.

Liker and Fleischer conducted a number of in-depth interviews with personnel from two divisions of a large automobile company that implemented DFM. Design for manufacturability was more difficult to introduce in the division that produced an existing product than in the division that developed and produced a new product and process. Liker and Fleischer identified several barriers to implementing DFM effectively. For example, career paths and rewards tended to discourage personnel from remaining in their positions long enough to learn across projects as well as to encourage conflict between functions as well as between hierarchical levels. In the latter case, conceptual design was performed at the corporate level, and detailed design was performed at the plant level. Effective DFM also requires changes in relationships between companies and their suppliers; however, the company's existing organizational structure encouraged conflicts of interest between these two parties that inhibited them from collaborating in their mutual long-term interest.

Francis and Winstanley focus on barriers to effective DFM within the context of British industry. British engineering has a strong craft and apprenticeship orientation and a strong functional focus. These conditions are incompatible with organizational practices that facilitate DFM. However, a number of social, economic, and technological changes are underway that may foster conditions that are more favorable to DFM or at least may "unfreeze" the present situation. Technological change is increasing the education level and professionalization of engineers and, at the same time, is undermining the craft and apprenticeship system; tough economic conditions are leading to contracting-out and the debureaucratization of organizations; and, finally, shorter product life cycles are encouraging the use of flexible working patterns. Organizational culture also may need to change in order to reinforce and maintain these new practices.

SUMMARY

This chapter discussed competitive strategies that companies can pursue in the emerging competitive environment as well as the strategic capabilities that these companies will need in order to achieve their objectives. The role of DFM in supporting these competitive strategies and, in particular, its role in facilitating the management of the product development process were discussed also. Finally, the social, political, and cultural context necessary to support and maintain DFM was reviewed. We will return to a strategic perspective on DFM at the end of the book, after the reader has gained an understanding of the issues that are raised in the following chapters.

NOTE

The terms design for manufacturability, design for manufacturing, and design for manufacture are used interchangeably throughout this book.

REFERENCES

Crosby, P. B. *Quality Is Free*. New York: New American Library, 1979.

Gomory, R. E. "From the 'Ladder of Science' to the Product Development Cycle." *Harvard Business Review* 67(6), November-December 1989, 99–105.

Kingdon, Donald R. *Matrix Organization: Managing Information Technologies*. London: Tavistock Publications, 1973.

Schonberger, R. *Japanese Manufacturing Techniques*. New York: The Free Press, 1982.

———. *World Class Manufacturing*. New York: The Free Press, 1986.

Stalk, G., and T. M. Hout. *Competing Against Time*. New York: The Free Press, 1990.

Susman, G. I., and J. W. Dean, Jr. "Strategic Use of Computer-Integrated Manufacturing in the Emerging Competitive Environment." *Computer-Integrated Manufacturing* 2(1), August 1989, 133–138.

I
STRATEGIC CAPABILITIES

2

COMPETITIVE ADVANTAGE THROUGH DESIGN TOOLS AND PRACTICES

STEPHEN R. ROSENTHAL and MOHAN V. TATIKONDA

Selecting and implementing design tools and practices for use in the development of new products is more of a corporate bet than a guaranteed investment. Driven by the external forces of market opportunity, technological progress, and competitive pressure, manufacturing companies typically review available design tools and practices, hoping to choose those that will help produce a sustained competitive advantage. Unfortunately, such choices are often accompanied by confusion, unreasonable expectations, or misdirected effort.

This chapter presents a conceptual framework that views design tools and practices in terms of promoting strategic capabilities that can become a source of competitive advantage. We define a competitive product to be one that stands the test of the market. It is, at worst, generally equivalent to competitor products, or at best, better than competitor products in one or more dimensions of interest to the customer. Such dimensions may include the following:

- Low product, system, and life-cycle cost
- Adequate or advanced functionality
- Integration capabilities and upgradability
- Convenient product availability in terms of quantity, location, acquisition time, and installation
- High reliability and non-intrusive serviceability
- Aesthetic and ''human'' fit of the product in its environment
- Confidence in the product

One or more of these customer expectations may receive top priority in any particular situation. For any given product, all may not be achievable at high levels.

We use the collective term *design tools and practices* to mean the approaches employed, either in a singular or an integrated fashion, to improve a product through its design and manufacture. These approaches can be methods, models, or even computer software packages used by those engaged in product design and development activities. Product improvements include lower unit cost, higher functionality, shorter manufacturing cycle time, shorter new product development (NPD) cycle time, lower

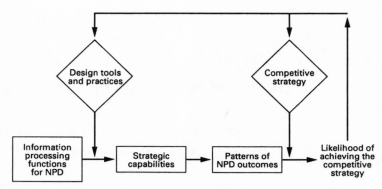

Figure 2.1. The link between design tools and practices and competitive strategy.

NPD cost, and higher end-product variety. These measures of product improvement all have something to do with the manufacture or manufacturability of the product. Note that some product improvements—such as shorter new product development times—might be transparent to the customer, yet integral to an organization's competitive strategy.

Figure 2.1 shows the set of linkages of interest in this chapter. This dynamic set of relationships centers on two kinds of decisions: the adoption of particular design tools and practices and the selection of a competitive strategy. It is desirable that the former support the latter even though their relationship is rather indirect. The connecting links are as follows. First (and of central concern in this chapter), certain information-processing functions serve to support these strategic capabilities for new product development. Second, design tools and practices can become potent managerial mechanisms for strengthening these strategic capabilities. Finally, the likelihood of achieving a particular competitive strategy is increased when a compatible pattern of NPD outcomes can be achieved over time (as distinguished from achieving the desired outcome from only a single new product). Then, as experience in new product development accumulates, competitive strategies may be modified and new design tools and practices may be pursued.

To explore these aspects of product design, we first consider three competitive strategies and the capabilities that facilitate their achievement. Then we identify and describe the six information-processing functions that our research points to as major contributors to successful new product development. Next, we further illustrate the competitive significance of these information-processing functions in terms of two important tools/practices: design for assembly (DFA) and quality function deployment (QFD). We conclude with a brief discussion of managerial implications.

Accordingly, when considering the adoption of any particular design tool or practice, management should encourage an objective evaluation of the type of information processing functions that it would support. A single design tool or practice may strengthen anywhere from one to all six of the functions, each of which will tend to improve certain aspects of new product development. If these improvements are critical enough, and if they are sustained over several successive new product introductions, the design tool or practice will have led to the creation of a continuing strategic capability. Certain capabilities, as outlined in this chapter, can be critical in promoting particular competitive strategies.

COMPETITIVE STRATEGIES AND SUPPORTING STRATEGIC CAPABILITIES

As mentioned in Chapter 1, Susman and Dean (1989) offer three strategies for competitive advantage that rely on skills in new product development. We review these strategies here as a starting point for identifying related strategic capabilities that, in turn, may be achieved through design tools and practices.

The first strategy is to compete on multiple dimensions, including low cost, high quality, and short delivery lead time. The new product development process deals most directly with the first two dimensions and less directly affects the third as well. Even when they are all deemed to be important, these three dimensions may have to be traded off to some degree, as particular design options for the product and associated production processes are identified and analyzed. Achieving this strategy calls for rigorous ways of identifying and evaluating the various design options. Management of the overall design process will also affect the likelihood of achieving this strategy.

A second competitive strategy is to compete in multiple segments. Here an attempt is made to gain a premium price in a number of different market segments by offering products that are highly differentiated to meet the needs of each of those segments. Especially important in the pursuit of this strategy is effective market research that is closely integrated with other functional roles in the NPD process. A great deal of customer involvement also may be beneficial throughout this product design and development process. Achieving the necessary market orientation to succeed in this strategy requires extraordinary commitment to enhanced communication, both human and computer-based. The costs of variety can be particularly troublesome when following this strategy unless products can be designed in a modular fashion or in terms of product families, with special attention to keeping the number of parts and suppliers as low as possible.

A third competitive strategy is to compete by continuous product improvement. An important prerequisite to this strategy is a product design and development process that is efficient, effective, and well-understood by all involved. Organizational learning about product development is also particularly important here to generate continuously improved NPD capabilities. The introduction of new products must become a seamless, ongoing business activity for companies pursuing this strategy. This strategy transcends the capabilities inherent in any single NPD effort, by requiring the maintenance of a regular rhythm of new product introductions (often coupled with short product development cycle times).

Because this strategy is based on close monitoring and appreciation of changes in customer desires, it also requires appropriate market research and ongoing forms of customer interaction, both carefully integrated with the NPD process. Also important is the ability to conceive of product families. In many industries this requires the ability to combine patterns of ongoing incremental improvements in technology with occasional carefully selected and well-timed technology "jumps." Flexible manufacturing capabilities can be essential to accommodate rapid and frequent product changes.

These three competitive strategies are dependent on at least two strategic capabilities that are associated with effective new product design and development:

- Cross-functional integration. This capability becomes a critical basis for solving difficult design problems and making essential trade-off decisions while retaining a focus on customer requirements.
- An efficient and effective NPD process. This capability facilitates, for example, the use of existing design data, appropriate development and testing approaches, and a flexible mode of manufacturing.

An effective product design strategy facilitates the development of these strategic capabilities through a compatible combination of design tools and practices, organizational structures, human resources, and managerial styles. Accordingly, particular design tools and practices need to be assessed in terms of their potential value towards achieving such capabilities. To some extent, both of these capabilities are needed to support each of the three competitive strategies outlined above, although the relative significance of particular steps will vary from strategy to strategy.

Our field-based research across several different industries suggests that these strategic capabilities can be developed by enhancing key information processing functions associated with product design and development. As shown earlier in Figure 2.1, information-processing functions, in turn, affect not only the outcome of a single NPD effort, but also the ability to sustain a successful pattern of NPD outcomes. It is this pattern that either fits or does not fit the company's competitive strategy. This descriptive model becomes useful only when the key information-processing functions of NPD are understood and incorporated into assessments of different design tools and practices.

INFORMATION PROCESSING FUNCTIONS IN PRODUCT DESIGN AND DEVELOPMENT

Most NPDs require the execution of a characteristic set of collective efforts. While different companies and industries may name them differently, the content is generally the same. We use the following terminology in this chapter to indicate the core phases that need to be integrated in a successful new product development project:

- Concept development
- Product design
- Prototype development and testing
- Process design and development
- Production ramp-up

Regardless of the nomenclature for the phases, product development is rarely a streamlined, efficient process. Several iterations are often required within phases. Phase activities are often overlapping, and even within one phase different functional tasks (for example, marketing, product engineering, and manufacturing) may occur in parallel. While new product development is often described as a set of functional tasks occurring in a sequential fashion, the process typically lacks the linearity suggested by this model (Clark and Fujimoto, 1989). Even for the development of relatively simple products, new product development is usually a complex organizational effort. As evidenced by the large number of failures in new product development, tasks within each phase or across multiple phases are often unintentionally skipped or inadequately executed.

All phases of the design and development of new products require that groups of functional specialists make a large number of decisions. Such decisions include the form, fit, and function of the product and how to manufacture it. In making these decisions, participants conduct design-related activities, some of which come to be quite routine and others of which are comparatively unique. Both types of design activity can be enhanced through the adoption of new tools and practices.

The selection, design, and development of new products is information intensive.[1] It requires accumulation of data and insight from diverse sources and different functional perspectives. As a new product is being designed and developed, information is constantly accessed, interpreted, augmented, transformed, and deployed. The accuracy, consistency, and availability of such information essentially determines the extent to which a new product achieves required functional capabilities, ease of manufacture, and a fit with overall strategies. Adopting an information-processing perspective on new product development can facilitate the strategic assessment of different design tools.

Design tools and practices can fundamentally enhance information processing and, therefore, problem-solving by new product development teams and related managerial groups. Aggregations of these design approaches can be chosen to provide ongoing strategic capabilities that can serve as a basis for enduring competitive advantage.

Priorities in the pursuit of a competitive edge through design tools and practices are suggested by the earlier discussion of competitive strategy. However, the ordering of priorities ought to depend on an organization's existing competencies as well as its competitive strategy. The most common pitfall is to try to adopt a complex tool or unfamiliar practice with inadequate appreciation for the magnitude of the undertaking. Given all of these contingencies, it is not surprising that no single implementation sequence seems to prevail. Our purpose in this chapter is to identify connections between design tools and practices and the achievement of elements of a competitive NPD process, rather than to provide a specific, all-purpose, "best practice" map for action.

Based on a recent study at Boston University of product design and manufacturing management,[2] we have identified a set of functions that are central to the successful design and development of new products. Each of these six information processing functions signifies a different form of information activity central to the product design and development process:

- Translation
- Focused information assembly
- Communication acceleration

Table 2.1. The Six Information-Processing Functions and Related Design Tools and Practices

Translation
Quality function deployment (QFD)
Design for assembly (DFA)
Customer use into test requirements
Target costs into yield objectives
Computer-aided process planning (CAPP)
Planning bills-of-material (BOM)
Value engineering

Focused information assembly
Early vendor involvement
Early manufacturing involvement
Simultaneous engineering
Co-location of design and manufacturing engineers
Quality function deployment (QFD)
Design for assembly (DFA)
Design reviews
Manufacturing systems simulation

Communication acceleration
Computer-aided design (CAD)
Group technology (GT)
Electronic data interchange (EDI)
Early specifications to vendors
Computer-integrated manufacturing (CIM)
Planning bills-of-material (BOM)
Preliminary prototypes
Rapid prototyping
Early product information to field service
Early product information to marketing/sales

- Productivity enhancement
- Analytical enhancement
- Management control

By paying more explicit attention to each of these functions and ways of achieving them, companies can develop an approach to pursuing their competitive strategy through the adoption of particular design tools and practices.[3] Below, we explain the significance of each of these functions and provide illustrations as to how they may be achieved.

Table 2.1 summarizes the predominant relationships between an illustrative set of design tools and practices and the six information-processing functions. The citations in this table could be expanded to include links of lesser significance than the ones shown and discussed below. Furthermore, other design tools and practices could be analyzed in a similar manner. In short, we have included an important set of relationships in this table and the associated discussion, but this should not be taken as a fully exhaustive analysis.

Our major theme is that a design tool or practice, successfully implemented, can strengthen one or more of the six information-processing functions. Depending on its context of use, configuration, and form of implementation, a given tool or practice can support different information-processing functions at varying levels. The information-processing functions that are strengthened by each of two particular de-

Table 2.1. *(continued)*

Productivity enhancement
Computer-aided design (CAD)
Computer-aided software engineering (CASE)
Project evaluation review technique (PERT)
Computer-aided engineering (CAE)
Group technology (GT)
Computer-aided manufacturing (CAM)

Analytical enhancement
Manufacturing simulation
Learning curve analysis
Computer-aided design (CAD)
Finite element analysis
Robust engineering
Statistical design of experiments
Taguchi methods
Design for assembly (DFA)
Quality function deployment (QFD)

Management control
Gantt charts
Project evaluation review technique (PERT)
Contract books
Formal performance reviews
Milestone gate reviews
Design for manufacturing (DFM) checklists
Manufacturing sign-offs
Group sign-offs

sign tools and practices are illustrated in a later section. As will become clear from that discussion, there are complementary groupings of tools and practices that strengthen each of the information-processing function categories.

Translation

The new product development process inherently requires that various kinds of specialized functional work be performed. A coherent NPD effort must assure that these specialized planning, design, and development activities are mutually consistent and compatible. This can be encouraged, particularly in a small company, by making sure that the various specialists interact regularly. Even then, the NPD effort is likely to be hindered by the inherent breakdowns in communication brought about by the isolated "thought worlds" that typically dominate the problem-solving orientation of individual functional specialists (Dougherty, 1989). What is needed at such times are design tool and practices that facilitate the transformation of sets of information from one point of view to another. This information processing function, which we call "translation," should be explicit and as routinized as possible.

For example, marketing product planners aim to determine customer desires for new products. These desires are often communicated in general terms that have little to do with implementable product capabilities. Use of quality function deployment techniques would facilitate translating, for example, "nice writing flow" for a pen to "xx viscosity of pen ink and yy roller ball pressure in pen." Subsequent to this marketing-to-product-design translation, a company might employ a design for assembly algorithm to translate product design specifications to particular manufacturing engineering requirements. Similarly, projected variations in usage environments for some new products need to be translated into specifications for various in-line test equipment. Target unit costs for a new product need to be translated into yield objectives for manufacturing ramp-up. Similarly, computer-aided process planning (CAPP) can translate product designs into manufacturing routings and detailed process plans. Planning bills-of-material (BOM) may translate a preliminary product part hierarchy to purchasing requirements and assembly facility layout. Value engineering can translate product functionality and service requirements into product cost and materials guidelines.

As suggested by these examples, the translation function is important both within and across the various phases of the NPD effort. In some instances translation is accomplished by a tool or technique that, once implemented, becomes an automatic part of ongoing NPD processes. Other translation practices, more dependent on the informal efforts of individual NPD participants, may need constant managerial reinforcement if this function is to have ongoing strategic significance. Since many of the "communication problems" in NPD can be traced to failures in translation, the leverage that can be achieved from building a powerful portfolio of tools and practices in this area is enormous.

Focused Information Assembly

At various points in the NPD process, certain design tools and practices may bring novel combinations of information to bear on a particular problem, thereby leading to greater choice and improved decisions. Often, this may be accomplished by routinely providing certain information to upstream stages earlier in the NPD process than otherwise would occur. Another type of improvement along the same line is to assemble sets of related information that are traditionally available at the same time but not in a consolidated form. This facilitates easy access and use. In either of these manifestations, we call this information-processing function focused information assembly.

The practice of involving vendors early in the NPD process leads to earlier knowledge of vendor capacities, capabilities, costs, and constraints. Another example of upstreaming is having manufacturing involved in early phases of the NPD process, thereby encouraging the use of existing parts and effective purchasing relationships. As discussed below, both quality function deployment and design for assembly techniques provide a framework for collecting and integrating multiple sources of information. Similarly, design reviews by groups composed of diverse technical and other specializations provide either formal or informal occasions for focused information assembly; such practices serve to aggregate varied and innovative perspectives on many aspects of design, tooling, and service. Manufacturing systems simulation requires collection and integration of detailed product and process information; this technique can provide valuable feedback for the improvement of both product and process designs. Activities of these types may also support the NPD effort by providing new awareness of existing information and more extensive usage of it.

Focused information assembly is especially valuable in support of non-routine design activity where participants with different functional backgrounds must work together. Such activity is often enhanced when novel interpretations and decisions arise owing to the mix of people involved. Here, synergies in interpretation arise from collaborative consideration of issues (more good ideas). For example, a colocated designer and manufacturing engineer, or such a tandem co-located by virtue of CAD and computer-integrated manufacturing (CIM) technologies, can simultaneously apply two broad but often different sets of knowledge to a problem. In this case, the manufacturing engineer brings knowledge of manufacturing equipment capabilities and product material trade-offs relative to manufacture, while the design engineer brings knowledge of desired product functions and general material abilities to achieve those functions. They may simultaneously optimize design and manufacturing but, often more importantly, bring whole new ideas and approaches to resolution of the problem at hand. Focused information assembly thus leads to greater choice and improved decisions.

Through the focus of multiple sources of knowledge, collaborative decision-making may lead to avoidance or resolution of problems that would otherwise not be resolved at all. Admittedly, it may be difficult to establish a workplace culture where effective collaboration in NPD is widespread (Rosenthal, 1990a). Nevertheless, one prerequisite for such collaborative efforts to succeed is that there be effective technological capabilities for the focused assembly of information.

Communication Acceleration

The timing of access to information is particularly critical to NPD success because it can directly affect the cycle time to introduce a new product. Design tools and practices that allow downstream stages of the process to start earlier (such as simultaneous engineering) may also reduce chances of oversight and error (and thereby reduce variability in the NPD process), and allow evaluation and selection of greater numbers of alternative approaches to the particular task. Such tools and practices achieve the function of communication acceleration.

An example of communication acceleration in the context of simultaneous engineering is the use of CAD coupled with group technology and electronic data interchange (EDI). This set of tools and practices helps vendors start early design of parts and tooling, reserve production capacity, and order materials earlier. Such computer-integrated manufacturing (CIM) technologies can greatly enhance the communication across functions. Similarly, the provision of a tentative product bill-of-materials assists manufacturing planners in formulating human assembler technical skill requirements and in turn supports hiring and training, all accomplished earlier than normal.

The development of early product prototypes, even in the concept development phase of an NPD effort, is another example of Communication Acceleration. Here, when design engineers provide an initial set of product specifications, the development of a preliminary prototype is made possible even before specifications are refined or formally approved. Also, technologies allowing early, rapid prototype development ensure that a physical unit or part is available early to those who will work on other downstream activities (such as tooling or packaging). Conducting development activities earlier than normal can yield insights and ideas valuable to the ongoing design effort.

Communication acceleration practices can also facilitate the portions of new product introductions that are carried out by field service and marketing/sales groups. Field service needs to train field engineers in new methods and perhaps develop new diagnostic equipment. Marketing/sales needs to develop promotional material and educate key actors in the product distribution network. Providing appropriate product design information to these groups as soon as it is available will allow them to get an early start on activities that are critical to a successful product release.

A common benefit of communication acceleration tools and practices is the potential for early development of needed skills and experience. Consider, for example, what can be done when manufacturing engineers are given early access to information on the production of a new product by participating in assembling engineering prototypes. This early production experience can lead to the development of better manufacturing tools, thereby reducing the extent of manufacturing start-up problems. Similarly, early knowledge of part specifications allows vendors to start production early and thus proceed down the learning curve. In both of these examples, communication acceleration ends up supporting quicker ramp-up to full volume production.

It should be clear from these examples that communication acceleration contains an element of risk not present in the information-processing functions of translation or focused information assembly. Here, in the interest of reduced product development cycles, a company may intentionally release information earlier than prudence would dictate. The goal is speed to market, which can have tremendous competitive significance. The cost, however, may be wasted effort from such "early starts." In general, the pursuit of reduced product development cycle times is likely to be sufficiently important to make communication acceleration an important function for NPD (Stalk and Hout, 1990). However, managers need to appreciate that the potential risks—in terms of increased product cost and reduced quality and market acceptance—argue against the single-minded pursuit of speed (Rosenthal and Tatikonda, 1990). Furthermore, managers must not confuse a strategy of competing through the continuous improvement of products with a strategy of increasingly accelerated NPDs. The former has to do with the rate of succession between NPD cycles, while the latter deals only with the speed of executing the NPD process.

Productivity Enhancement

The efficiency with which some required design or development task is conducted can be improved directly by adopting certain design tools and practices. Such enhancement of NPD productivity can be significant for its own sake, independent of improvements in the other kinds of information-processing functions already described. We define these productivity enhancements to be the use of tools or practices that improve the speed or reliability with which one or more of the routine activities of new product development takes place. Much as productivity has long been viewed as an objective in the field of manufacturing, this same perspective can be applied to the process of product design and development. Admittedly, the narrow-minded pursuit of productivity improvement in this arena can raise strategic dilemmas similar to those in the field of manufacturing, such as the loss of flexibility through excessive refinement of tasks, roles, and equipment (Skinner, 1986). Nevertheless, it would be a mistake not to acknowledge and selectively pursue the productivity payoffs that are available.

A drafting-oriented CAD package, for example, allows for quicker (relative to traditional manual methods) retrieval, drawing, and redrawing of parts and schematics. The CAD package would also provide other related and required documentation for manufacturing and purchasing purposes, tasks that if not computer supported would have to be done by hand. This tool not only speeds up accomplishment of certain required tasks, but also reduces errors and supports more consistent and reliable information. Likewise, computer-aided software engineering (CASE) techniques can improve the productivity of the software designer, who is increasingly a key participant in the development of a wide variety of products and systems. At the broader level of project planning and scheduling, critical path methods and techniques (e.g., project evaluation review technique [PERT]) may facilitate enhanced productivity by the product development team, and also reduce project planning and related management efforts. Computer-aided engineering (CAE) allows quicker rudimentary testing and analysis of proposed designs. Group technology in its many forms supports retrieval of old product and process information, reducing the need to "recreate the wheel." Computer-aided manufacturing (CAM) equipment that automates robotic and numerical-control programming activity can save time and minimize coding errors.

Analytical Enhancement

Certain analytical tasks are conducted as a matter of course in any traditional new product development project. Some design tools and practices enlarge the feasible analytical range by providing unique capabilities to support the work of design engineers, product strategists, or other participants in the NPD process. Such analytical enhancements are akin to those achieved by manufacturing tools such as job shop simulations or learning-curve analyses, which provide higher-level, sophisticated information not available before. Similarly, these are applied to the design of new product or process capabilities. Tools for analytical enhancement make it possible to assure in advance that the new product will be more consistent than otherwise with some of the requirements of customers.

A CAD software package for conducting finite element analysis, for example, allows designers to consider thermal gradients and other physical properties to a degree that might not be possible otherwise at the early stage of designing a new product or part. Results from such analysis increase the effectiveness of the new product development process, and can lead to reduced product cost, greater product reliability, and higher functionality.

Design engineers are also charged with developing "robust" product designs, which enable the product to perform its intended function well, even in extremely unfavorable environmental contexts and operational modes. By using statistical design of experiments, simulation and optimization models, and other statistical methods, design engineers can identify design parameters that cause product performance to change very little despite a wide range of potential environmental and use conditions. These statistical techniques in aggregate are often called Taguchi methods. The practical payoff from this analytical enhancement is that the other, more influential, or sensitive, parameters are assigned appropriate design and manufacturing tolerances, thereby improving the resulting product reliability.

Management Control

Certain design tools and practices facilitate the assessment, control, and evaluation of the NPD process, either in its entirety or with respect to specific issues. We call this the management control function because it serves to ensure a more thorough and systematic NPD process. Some control tools, like Gantt charts, PERT charts, and other kinds of project plans, help ensure that minor and major steps in the process are not neglected, guaranteeing greater process completeness and adherence to sequence. Other techniques, such as a "contract book" specifying at an early stage of the NPD which organizations will deliver on specific aspects of the new product and associated production process, facilitate the use of common targets and a shared NPD vision.

Other practices, such as formal performance reviews, reduce the risk that is inherent in any new product introduction effort. Along these lines, many companies now use a "stage/gate" system of performance reviews. Following this practice an executive review team will examine the NPD with an eye toward passing key milestones such as concept approval, product design release, design freeze, and manufacturing release. Satisfying the executive review team that all issues have been adequately addressed (and problems solved) at each of these stages is a prerequisite for passing through this gate and proceeding to the next stage of the NPD. Such aids to management control may promote both the efficiency and the effectiveness of new product introductions.

Where product cost is a serious basis for competition, one has an incentive to establish procedures and methods for confirming that materials and manufacturing costs are consistent with early assumptions. This includes attention to special equipment, tools, and fixtures. Where product performance is especially important, demonstrations at various stages of the development process (e.g., engineering prototypes and pilot runs), coupled with state-of-the-art test capabilities, are common. Assessing the designs for test equipment is particularly critical in those industries where the test equipment can be more complex than the product that needs to be tested. Formal procedures for the qualification of key suppliers are also important in a growing number of industries where conformance, dependability, and reliability requirements are reaching new heights. Design for manufacturability checklists, requiring design and manufacturing engineers to consider carefully each part for manufacturability, and manufacturing sign-offs on product designs to guarantee manufacturing's willingness to accept responsibility for producing the product are also in widespread use.

Practices of this sort are a necessary but costly and time-consuming part of controlling the progress and outcomes of an NPD project. The challenge is to establish a cost-effective set of managerial controls (Rosenthal and Tatikonda, 1990). While good news (no problems) is always desirable, the real purpose of such approaches is to generate bad news (anticipation of design or development problems) as early as possible.

Synthesis

The six information-processing functions described above can be thought of as the life support systems of a company's new product development capability. Regardless of the complexity of the product, the size of the company, and the number and skill level of its people, these functions are all essential. The six functions can be grouped into two sets, each promoting a different type of capability: (1) cross-functional integration and (2) an efficient and effective NPD process.

Cross-functional integration is accomplished through translation, focused information assembly, and communication acceleration. The common theme across these three information-processing functions is furthering NPD communication in ways that link related streams of design and development activity. The translation function accomplishes this by converting critical information from the viewpoint of one specialist to that of another. Focused information assembly brings together, in both time and place, information that supports improved design and development decisions. Communication acceleration operates in the time dimension, bringing vital information to NPD participants sooner than they would otherwise have it. This set of functions reduces the level of "noise" in complex networks of design and development decisions. It promotes a common vision for the NPD effort as a whole and it strengthens the connections across the contributions of individual participants and sub-teams.

There may be some trade-offs among these three cross-functional integration functions. For example, a strong emphasis on focused information assembly will occasionally delay the timing of certain key decisions that, in turn, may offset other gains in project timing resulting from initiatives in communication acceleration. There are likely to be some synergies as well. For example, focused information assembly coupled with communication acceleration may facilitate more rapid skill development and smoother transfer from design engineering to manufacturing: many companies are creating initial production organizations, combining both design and manufacturing engineers, with this goal in mind.

An efficient and effective NPD process is accomplished through productivity enhancement, analytical enhancement, and management control. "Efficiency" deals directly with the expenditure of time and resources of all types in the NPD process. "Effectiveness" refers to the level of success of the NPD project in introducing a product that meets specifications set for it. Productivity enhancement and analytical enhancement contribute to efficiency and effectiveness through the introduction of new tools and practices—productivity enhancement through the reduction of costs (often by using automation) and analytical enhancement through the achievement of desired product capabilities. Management control practices also promote NPD efficiency and effectiveness, but in a less direct manner, by establishing information bases, procedures, and structures for more constructive review and control over the NPD process. Design tools and practices associated with these three information-processing functions can be justified in terms of specific contributions that they make to the efficiency and effectiveness of the NPD process.

There may be a critical trade-off between productivity enhancement and management control. If attempts at management control get overly bureaucratic, the associated loss of time can easily offset any gains received through productivity enhancement tools and practices. The opportunities for synergies across these functions are also significant; tools and practices for management control can promote the adoption of tools and practices for both productivity and analytic enhancement.

By contrast, the three types of integration functions affect NPD efficiency and effectiveness in less transparent ways, as they promote smoother and more tightly linked interpersonal activities. Admittedly, integration functions can be seen in retrospect to improve NPD aspects of efficiency and effectiveness, but they are not so clearly associated with such outcomes in advance.

ANALYSIS OF TWO SELECTED DESIGN TOOLS/PRACTICES

Below we describe and analyze design for assembly (DFA) and quality function deployment (QFD). We have selected these two techniques for special treatment because they are particularly rich in their possible impacts on new product development. Each of these is being actively considered for widespread adoption in different industries. As a basis for understanding them better, we summarize their particular methodologies and applications, and their traditional expected NPD outcomes. We then identify information-processing functions that they tend to strengthen in a dominant way and indicate how this promotes particular sustained strategic capabilities.

Design for Assembly

Design for assembly (DFA) is a systematic analysis process primarily intended to reduce the assembly costs of a product by simplifying the product design. It does so by first reducing the number of parts in the product design and then by ensuring that remaining parts are easily assemblable (Boothroyd and Dewhurst, 1987). This close analysis of the design is typically conducted by a team of design and manufacturing engineers, although other functional expertise such as field service and purchasing may also be included. DFA is used for discrete manufacturing products, and primarily for durable goods, but occasionally for consumer products. Since it is used to optimize assemblies, it is often used for smaller and medium-sized products, or for many sub-elements of larger systems. DFA does not specifically support system level applications and is usually applied to subassemblies (Stoll, 1988).

Whether conducted manually or with software, DFA techniques lead to a simpler product structure and assembly system. DFA algorithms build on many earlier industrial concepts, including group technology, producibility engineering, product rationalization, and time and motion studies. In many ways DFA is a structured, automated approach to time and motion industrial engineering, combined with a bit of design philosophy via design axioms and guidelines (Andreasen et al., 1983).

There are two uses of DFA. It may be used to redesign a product already in manufacture (or a product being "remarketed," or reverse-engineered). In this case, the product is disassembled and reassembled with special consideration of parts handling (feeding and orienting) and attachment (insertion) times and costs. These times and costs are found in data tables, through software, or by empirical observations. DFA also may be used for analysis of a product while it is in design.

DFA was developed with the assumption that a significant portion of manufacturing costs is set in the design stage itself, before any manufacturing systems analysis and tooling development are undertaken (Nevins and Whitney, 1989). The primary objective of DFA is to minimize parts counts, thereby having fewer parts to be manufactured and assembled, fewer parts that can fail, and fewer interfaces between parts. It is these part interfaces that contribute primarily to product failure by providing sources for failure (Welter, 1989). The second objective is to have remaining parts that can be easily assembled (Boothroyd and Dewhurst, 1987).

DFA provides a quantitative method for evaluating the cost and manufacturability of the design during the design stage itself. Some firms may choose to apply the analysis later in the design stage, and others may do so quite early, such as during initial concept evaluations. In either case, the analysis steps require that an initial design be developed or proposed first. Then this design alternative is assessed penalty points for each undesirable feature of the design. These points when aggregated help determine the "design score" efficiency of assembly, based on the expected material and assembly costs. Then the product is "redesigned" (this can be conceptual only) using part and product level design rules coupled with consideration of annual volumes and existing manufacturing processes. This step requires engineering creativity because DFA, even with rules, guidelines, and measures, is still an art. A typical design guideline is achieved by software queries asking these questions in the case of two parts connected by a fastener: Does the fastener part move? Does it have to be a different material from the two parts? Does it have to be removed for servicing? If the DFA team's response to all three questions is no, then the software would advise the team to make the assembly as a single part (Andreasen et al., 1983).

DFA has also served as a tool for supplier selection and involvement. This aspect is highlighted as large firms continue to move design and assembly functions to suppliers. Ford requires certain vendors to evaluate their designs with DFA before they can submit bids (Kirkland, 1988). DFA can also provide leverage over suppliers since firms can estimate product costs better.

Depending on how it is used, DFA can serve one or more of the following information processing functions described in the previous section. To begin, DFA facilitates *translation* from product design configuration to assembly cost, exact parts, and resultant equipment and personnel needs. It also provides important *analytical enhancement* because the DFA algorithms enable new kinds of subassembly design analysis that precede the development of physical prototypes and generate early findings that might otherwise not arise.

DFA serves other information-processing functions in a more secondary manner. Its relationship to *focused information assembly* is that the practice of DFA requires the bringing together of particular product configurations, general assembly costs and times, design axioms, and manufacturing systems principles, all to support a design choice. When used in early design or prototype analysis, DFA provides early information to manufacturing and production, planning, and control regarding needed equipment, labor skills, and parts. In this fashion DFA supports the firm's *communication acceleration* function. Another contribution of DFA is in the realm of *management control:* this practice enforces desired reductions in parts count, provides a basis for resolving issues of manufacturability, strengthens supplier management, focuses engineering efforts, and provides a relative evaluation score for alternative designs.

The collective strategic impacts of DFA can be considerable. Design for assembly is both a tool and an organizational mechanism, or—in the terms of this chapter—both a design tool and practice. Since DFA requires teams of some sort, previous experience with cross-functional interaction, data collection, and analysis supports DFA use. Alternatively, DFA helps build such experience. DFA may be seen either as being equivalent to "concurrent engineering" or as a subset or a precursor to it. The organization that uses DFA improves and learns as many individuals work together concurrently collecting information, documenting development efforts, and applying improved bases for decision-making. In Boothroyd Dewhurst, Inc. promotional literature, a Xerox executive is quoted as saying that when engineers use DFA software "they become more proficient in understanding the product delivery process so that each design becomes better and better." Ford uses DFA as part of their simultaneous (or concurrent) engineering efforts, stating, "it goes a long way to break down the barriers that exist between groups" (Welter, 1989).

DFA requires that design and manufacturing engineers coordinate activities but does not specifically require the involvement of others. Its actual value in serving the information focus function may thus be less than is ideally achievable. For example, while IBM's Proprinter II is lauded for a fourfold increase in reliability, it is not considered to be easily serviceable; in this case, field service had limited involvement in the DFA team (Bebb, 1990). DFA can be used to integrate supplier involvement in the NPD process but has limited ability to involve customers (though some may be on DFA teams for usability input).

DFA increases the time and resources spent in the design stage of the NPD but cuts time throughout the rest of the development process. It does so by ensuring only one iteration of design and manufacturing, by reducing tooling acquisition and development requirements, by lessening training needs, by involving manufacturing and supplier functions earlier, and by providing the many time-saving benefits derived from simultaneity. DFA does not necessarily increase the very up-front product planning and system level configuration planning, but it does take more time when tangible subassembly design alternatives are in hand.

Quality Function Deployment (QFD)

QFD is a systematic approach for the design of new products or services based on close awareness of customer desires coupled with integration of corporate functional groups. This approach is a set of planning and communication techniques that focus and coordinate organizational capabilities in order to develop products that most closely meet customers' actual needs. QFD can be seen in two lights: (1) as a comprehensive organizational mechanism for planning and control of new product development or (2) as a localized technique to translate the requirements of one functional group into the supporting requirements of a downstream functional group. Most U.S. use is of the localized variety, and so we focus primarily on this type of application (King, 1989; Akao and Kogure, 1983).

QFD requires a multifunctional team of experts to participate simultaneously in a complex analytic exercise, bringing diverse perspectives to bear on key issues of design. In localized uses, QFD serves both technical and organizational functions. Primary expected outcomes of QFD are the following: (1) There is an increased awareness of the customers' desires; improved product with focus on actual customer wants; improved product specification setting; and, in general, more successful and effective products, (2) Synergistic gains result from integration of, and increased effective communication between, individuals from different functional groups, such gains include understanding each other's functional constraints and capabilities and transfer of information earlier than normal, (3) Effective definition and prioritization of new product development activities leads to increased understanding of trade-offs, clear understanding of product objectives and customer needs, improved decision making, a better end-product and reduced product development time (owing to elimination of remedial design iterations).

The most common example of QFD is translation from customer desires regarding improvement of an existing product into actual design engineering changes to be brought about in the next version of the product. This application is often referred to as the "house of quality" due to the shape of the data matrix used to accomplish the translation (Hauser and Clausing, 1988). Customer demands are determined in customers' own terminology via various market research techniques and are ranked in terms of relative importance. Such a demand might be "ease of writing flow" for a pen. Each design characteristic (such as pen ink viscosity or pressure on ball point) is then correlated with each customer desire to determine the degree to which it achieves the customer requirement. Certain design characteristics may complement or offset each other—these trade-offs and opportunities are noted. Customer perceptions of competitor's products are generated to scale the advantages and disadvantages of the firm's product. Priorities and target values are set in an attempt to satisfy customer desires through particular design characteristics (e.g., low ink viscosity and high ball point pressure). Another localized application might have product design characteristics translated into required manufacturing processes to support production of the product (Smith, 1990).

QFD revises the time and intellectual energy commitments at each stage of the NPD process. In traditional NPD projects, the concept development phase may be somewhat short, with long periods of time devoted to such activities as design, manufacturing implementation, and redesign iteration. QFD greatly increases the time and resource requirements during the early planning/conceptualization phase, but, if successfully implemented, reduces resource requirements for downstream stages.

The most common U.S. approach to QFD uses the house of quality method to trace customer requirements through to manufacture. This approach is quite effective for improving the cost, reliability, and functionality of existing product subsystems and parts. It is of limited use, however, for the design of entire systems or those with dynamic product concepts. This limitation in the localized QFD application is due to a dependence on existing company databases that reflect prior system level product concepts and familiar technological options for the product and the manufacturing process. Furthermore, American QFD does not generate new product ideas and is therefore frequently coupled with implementation of the Stuart Pugh (1981) new concept selection method. This method forces engineers to develop and evaluate many new concepts and so promotes fresh thinking. It is used as an input to the QFD process (Clausing and Pugh, 1991).

QFD serves to remove non-value-added activities by avoiding redesigns, by streamlining communication, and by reducing errors. It focuses energies on prioritized activities and so serves as a method for planning and control of design and other specific tasks, leading to more efficient and less variable processes. QFD can act as a mechanism to make sure options and steps are not forgotten, again reducing chances of downstream problems. Through QFD, creative energies are less likely to be squandered on activities that will not eventually benefit the product. Instead, QFD adds value to the product and NPD process by better coordinating information and people to better meet customer requirements and increase organizational capabilities.

QFD performs the *translation* function, in the form of requirements, first from marketing to product engineering and then to manufacturing. QFD also provides *analytical enhancement,* as it facilitates systematic data collection, comparison, and decision-making. Through the application of QFD, many experts come to understand both the new product and the new product development process to a much more complete degree. The customer voice is injected throughout QFD activities, and so engineers and others see design characteristics in a new light.

QFD serves to *focus information assembly* by encouraging unhindered and informal cross-functional communication, data sharing, and problem resolution that would not otherwise happen. A major side benefit of such team building is the avoidance of bureaucratic structures and barriers to communication. Because many of the participants in a QFD exercise normally would not participate in the NPD process until a later stage, this technique also acts, secondarily, as a *communication accelerator.* In addition, QFD serves a *management control* function, especially when implemented on a large scale, since alternatives and trade-offs must be explicitly considered, and decisions must be agreed upon by a group.

Sustained usage of QFD can lead to a number of beneficial strategic outcomes. Among them are faster new product time-to-market, early and/or faster starts for subsequent NPDs, availability of continuous supplies of new products to customers, and quicker compilation and updating of complete product lines. A favorable reputation for customer response will arise, as will significant product differentiation. On the organizational side, QFD leads to development and use of a high-effectiveness communications base within the company and supports organizational learning through enhanced documentation, sustained interfunctional relationships, and improved coordination. QFD is a foundation for "NPD rhythm," the special environment in which product development is a routine activity with a planned frequency. However, QFD may not support the frequent adoption of the newest product and process technologies because this technique is most rigorous when building on data from past product design and development projects.

SUMMARY AND CONCLUSIONS

This chapter has described how the adoption of new design tools and practices simultaneously shapes a design strategy with competitive implications. This will be true in general, independent of the particular competitive strategy being pursued. However, in comparing the three different competitive strategies outlined at the beginning of this chapter, some differences in emphasis can be noted.

For a company seeking to compete on multiple dimensions, prospective design tools and practices need to be assessed in terms of their ultimate support for product and process designs that promote the simultaneous pursuit of low cost, high quality, and short delivery time. Translation and focused information assembly between design engineering and manufacturing are likely to be particularly important information-processing functions in these situations. For a company seeking to compete in multiple segments, perhaps the most important information-processing functions are translation between marketing and other functions, coupled with an associated focus on the assembly of information. For a company seeking to compete by continuous product improvement, the speed and effectiveness of the entire NPD process are critical. Here, communication acceleration, productivity enhancement, analytical enhancement, and managerial control become particularly important (but not to the exclusion of the other two information processing functions).

This chapter has provided a conceptual framework to guide decisions on such technological and organizational innovations. The framework takes the point of view that the NPD process is itself a production system for new products and so should be considered from the viewpoint of production management. The six described information-processing functions are, in effect, critical components of a system to produce competitive new products through a process of design and development. One assesses changes to such a production system using the traditional metrics of elapsed time, and cost and quality of the output.

The framework developed in this chapter, as shown in Figure 2.1, thus has prescriptive implications. Starting with an existing or contemplated competitive strategy, a company can identify those strategic capabilities for new product design and development that best support the competitive strategy. Comparing the desired capabilities with the current status will lead to an appreciation for needed improvements in NPD-related information-processing functions. This provides a basis for evaluating the potential value in pursuing any particular design tool or practice. It also provides a framework for assessing the array of design tools and practices that are already in place. In other words, once the connections shown in Figure 2.1 become clear, management can reverse the logic and proactively assess design tools and practices in terms of the likelihood that they will help achieve the desired competitive strategy. More radical, perhaps, is the notion that—armed with such insights—management would do well to formulate competitive strategy with an eye toward the reality of their current and projected NPD strengths.

This line of thought requires attention to what we believe is an emerging dilemma: as managers place increasing attention on individual aspects of new product development (such as the development of cross-functional teams), there may be less of a strategic orientation to the set of elements that affect the overall process of product design and development. This would be unfortunate because it makes the individual initiatives less potent. How is this trap to be avoided?

We believe the solution requires concerted efforts of senior managers, strategic planners, and managers of product development. These three groups must modify and then combine their perspectives in ways that are not traditional in most large companies. In particular, senior managers need to go beyond competitive mission statements and the adoption of one or two favored design tools or practices. They should also promote a widespread awareness, through training programs and other devices, of capabilities that support the existing competitive strategy. Strategic planners, in turn, must learn to deal less in abstractions and more in terms of specific NPD capabilities and limitations. Finally, product development managers should take more deliberate steps to ensure that strategic benefits, not just tactical outcomes for a single product offering, accrue from investments in new design tools and practices. The framework presented in this chapter can be used to support a more methodical introduction of new design tools and practices along with the strengthening of existing ones. The common goal should be the development of a more coherent design strategy.

An agenda of this sort, given a high enough priority, should lead to needed reflection on the competitive advantage that can be gained through different combinations of design tools and practices. Together, these three groups of corporate actors with different natural orientations can bring about what is likely to be a radical change in thinking. Then it becomes more likely that others in such companies—those who work on portions of a new product development effort—will be able to do so with appropriate technology and an explicit strategic orientation.

ACKNOWLEDGMENT

The authors would like to thank everyone who contributed other chapters in this book for their thoughtful comments on an initial draft of the present chapter. We particularly appreciate the care and insight offered by Gerald I. Susman, the editor of this volume, in helping us shape our argument and presentations.

NOTES

1. This general perspective on product design as an information-intensive activity is not new. It has been presented in different ways in the D.B.A. dissertation of Takahiro Fujimoto (1989) and in some of his collaborative publications with Kim Clark, as well as engineering-oriented publications, such as the work of Nevins and Whitney (1989). This chapter is unique, however, in attempting to be more explicit in identifying the range of information-processing functions that are central to NPD, the ways in which design tools and practices support those functions, and the strategic significance of so doing.

2. This study, sponsored by the Boston University Manufacturing Roundtable, included the preparation of case histories of specific new product introductions in seven companies across different product types and industries. This study also included a series of workshops with members of these companies. Further description of this research project can be found in Rosenthal (1990b). Readers who wish to have additional information on other reports and working papers from this project may contact the Manufacturing Roundtable, School of Management, Boston University, Boston, MA 02215.

3. These six terms are not in current use in companies, although the underlying functions are familiar (at least in tacit form). We have formulated them inductively, by observing (across a number of firms and industries) what is being adopted and the kinds of gains being sought. In a more systematic survey, we would expect to find the most prevalent approaches to establishing these information-processing functions to differ by industry as well as by a company's strategy within its industry.

REFERENCES

Akao, Y., and M. Kogure. "Quality Function Deployment and CWQC in Japan." *Quality Progress* 16(10), October 1983, 25–29.

Andreasen, M. M., S. Kahler, and T. Lund. *Design for Assembly*. New York: Springer-Verlag, 1983.

Bebb, H. Barry. "Implementation of Concurrent Engineering Practices." *Concurrent Engineering Design Conference Proceedings*, Society of Manufacturing Engineers, Dearborn, Michigan, 1990.

Boothroyd, G., and P. Dewhurst. *Product Design*

for Assembly. Wakefield, R.I.: Boothroyd Dewhurst Incorporated, 1987.

Clark, Kim B., and Takahiro Fujimoto. "Overlapping Problem-Solving in Product Development." In *Managing International Manufacturing*, ed. Kasra Ferdows. Amsterdam: North-Holland, 1989, 127–152.

Clausing, Don, and Stuart Pugh. "Enhanced QFD." *Proceedings of the Design Productivity Institute International Conference*, Honolulu, February 1991.

Dougherty, Deborah. *Interpretive Barriers to Successful Product Innovation*. Report No. 89–114. Marketing Science Institute, Cambridge, Mass., September 1989.

Fujimoto, Takahiro. "Organizations for Effective Product Development: The Case of the Global Automobile Industry." D.B.A. thesis, Harvard University, 1989.

Hauser, John R., and Don Clausing. "The House of Quality." *Harvard Business Review* 66(3) May-June 1988, 63–73.

King, Bob. *Better Designs in Half the Time: Implementing QFD in America*. 3rd ed. Methuen, Mass.: Goal/QPC, 1989.

Kirkland, Carl. "Meet Two Architects of Design-Integrated Manufacturing." *Plastics World*, December 1988.

Nevins, James L., and Daniel E. Whitney, eds. *Concurrent Design of Products and Processes*. New York: McGraw-Hill, 1989.

Pugh, Stuart. "Concept Selection—A Method That Works." *Proceedings of the International Conference on Engineering Design*, Rome, 1981.

Rosenthal, Stephen R. "Bridging the Cultures of Engineers: Challenges in Organizing for Manufacturable Product Design." In *Managing the Design-Manufacturing Process*, eds. John E. Ettlie and Henry Stoll. New York: McGraw-Hill, 1990a, 21–52.

———. *Building a Workplace Culture to Support New Product Introduction*. Boston University Manufacturing Roundtable, September 1990b.

——— and Mohan V. Tatikonda. *Managing the Time Dimension in the, New Product Development Cycle*. Boston University Manufacturing Roundtable, August 1990.

Skinner, Wickham. "The Productivity Paradox." *Harvard Business Review* 64(4), July-August 1986, 55–59.

Smith, Larry. "QFD Implementation at Ford." *Concurrent Engineering Design Conference Proceedings*, Society of Manufacturing Engineers, Dearborn, Michigan, 1990.

Stalk, George, and Thomas M. Hout. *Competing Against Time*. New York: The Free Press, 1990.

Stoll, Henry W. "Design for Manufacture." *Manufacturing Engineering* 100(1), January 1988, 67–73.

Susman, Gerald I., and James W. Dean, Jr. "Strategic Use of Computer-Integrated Manufacturing in the Emerging Competitive Environment." *Computer-Integrated Manufacturing Systems* 2(2), August 1989, 133–138.

Welter, Therese R. "Designing For Manufacture and Assembly." *Industry Week*, September 4, 1989.

3

DESIGN FOR MANUFACTURING IN AN ENVIRONMENT OF CONTINUOUS CHANGE

SUSAN WALSH SANDERSON

The objectives of design for manufacturing (DFM) are to identify product concepts that are inherently easy to manufacture, to focus on component design for ease of manufacturing and assembly, and to integrate manufacturing process design and product design to ensure better matching of needs and requirements. Most research on design for manufacturing has focused on the interaction between design and manufacturing in the development of initial prototypes for new products and ignored the problem of upgrades, extensions, and enhancements to products that occur in the normal life cycle of a product. Design for manufacturing is often interpreted as ''get it right the first time'' with little if any concern for the management of the design process over time.

However, success in today's highly competitive environment depends not only on getting an initial product to market but in continuously upgrading and improving that product over time. Rapid change from one product generation to the next and a large number of models offered in any single year have proved to be a major challenge for manufacturers. Although American firms are learning how to do DFM on a project-by-project basis, most have been notoriously bad at carrying out DFM on a continuous basis. Many projects have a high degree of autonomy, and there is little continuity between one project and the next. The emphasis on shortening product development cycles may have even increased that tendency as project directors maneuver to control more and more of the resources they need to get their product out quickly. This shifting of a growing share of resources to independent projects may have even exacerbated the lack of continuity between projects.

U.S. firms that hope to compete in an increasingly global and competitive environment will need to bring a wider variety of high-quality, low-cost products to increasingly sophisticated and discriminating global markets that characterize everything from cars to computer chips. Current methods of product development, from design to manufacturing to marketing, cannot react quickly enough nor accommodate the diversity necessary to meet these new demands. New design technologies combined with new methods of engineering design management are needed to address these challenges.

There is growing anecdotal and empirical evidence that Japanese firms rely on their tradition of cooperation inside the firm and proximity and close relationships between firms to maximize the speed with which design and manufacturing are performed. While the Japanese have certainly made effective use of information technology and design tools, their real strength lies in their cohesive organizational relationships and the discipline they apply to the management of product families. They have followed a design management paradigm that emphasizes incremental innovation and careful management of product families. This management paradigm has important implications for the continuity of organizational teams and design practice and strongly influences the ability to undertake long-term strategic thrusts and react to changing market demands. By contrast, American firms have tended to emphasize more discontinuous product design focusing on independent model development and radical innovation.

Many U.S. firms are just beginning to overcome the internal separation between design engineering and manufacturing. For U.S. firms that have not matched Japanese-style relationships (both within and between organizations), emerging design and manufacturing technologies offer the opportunity to rethink traditional design management approaches. New technologies for design automation and flexible manufacturing may provide significant benefits, but their profitable use is strongly dependent on effective management and a more disciplined approach to design. If current and emerging tools are developed and implemented with more management insight than some of their predecessors, it is likely that firms will be able to achieve significant time, cost, and quality improvements and may be able to surpass, or at least keep pace, with the Japanese in the effectiveness of their design and manufacturing systems.

A STRATEGIC APPROACH TO DESIGN FOR MANUFACTURING

Effective design for manufacturing in today's fast-paced environment requires the management of a product family from early conceptualization in the initial product prototype as well as continuous improvement throughout the life cycle of the product. Firms at the leading edge of consumer markets create a series of models over the life cycle of a product family, advancing technical frontiers to lower costs and achieve a high degree of product differentiation (Sanderson and Uzumeri, 1990; Shintaku, 1990). In products ranging from televisions and VCRs to automobiles, they offer products that are attractive to customers in a wide range of prices and models.

Womack, Jones, and Roos (1990, p. 119) have shown that Japanese automobile firms have been able to expand their product range rapidly, nearly doubling their product portfolios from 47 to 84 models between 1982 and 1990. By contrast, European volume producers reduced their model offering slightly from 49 to 43 and kept models in the market longer. Although U.S. auto producers managed to increase substantially their product ranges from 36 to 53 models, they did so by keeping models in the market longer. The average age of U.S. models was 2.7 to 4.7 years as compared to 1.5 to 2.0 years among Japanese producers (p. 121). Womack and colleagues suggest that this is largely the result of the U.S. firms' inefficient product development processes, which require more engineers than comparable Japanese processes to expand product ranges and renew products (p. 121).

Firms that are successful in fast-paced product lines like automobiles and consumer electronics are masters at incremental innovation (Gomory, 1989). The inability of many American firms to renew and refresh their product lines, as well as keep up with the dizzying pace of technological innovation, caused most of them to exit the industry by the early 1980s. By contrast, Japan has a number of strong competitors matching each other's product offerings almost as quickly as they enter the market.

The most successful Japanese competitors in fast-cycle products combine great market awareness and sensitivity with excellent design and manufacturing. They have great discipline in spinning out successive iterations of products targeted to micromarket segments, and they perform periodic major engineering and manufacturing feats to move the product to a new technological frontier. They do so largely through an organizational commitment to continuous improvement with early attention to the development of clear concepts and goals for product families. Although many firms are engineering driven for the bulk of their development efforts, some firms have been able successfully to combine insight from marketing and industrial design to produce products with wide visual appeal to customers. Sony in consumer electronics and Toyota and Mazda in their high-end automobiles have produced products with considerable design appeal to complement their engineering excellence.

Managing Product Family Evolution

Two different product development paradigms have dominated the management of design and manufacturing in the United States and Japan. The first, which characterized much of product development until recently, can be termed ''revolutionary'' or discontinuous (see Figure 3.1). This pattern views product development as a series of largely unrelated models produced by independent development teams. A second paradigm, which we have termed ''evolutionary,'' takes a more planned and linked approach to new model development (see Figure 3.1). The second paradigm resembles the development of a tree, where the branches and leaves extend from larger limbs.

Sony has followed the evolutionary paradigm in the development of many of its families of consumer products. By rapidly upgrading and enhancing the original product concept as well as offering a wide variety of models targeted to different price segments, Sony, one of the premier consumer electronics firms in the world, has dominated personal stereo (Walkman) sales since 1979 (Sanderson and Uzumeri, 1990).

Sony has pursued a similar strategy in developing a number of other products— portable CD players and the 8-mm camcorder as well as other product families. A similar pattern is followed by such major successful Japanese competitors as Casio and Sharp in calculators (Shintaku, 1990). These firms offer a full line of models spanning a range of features that have evolved through successive regular innovations. This strategy has allowed them to win dominant market share (Sanderson and Uzumeri, 1990; Shintaku, 1990).

Sony has successfully used platforms from which to spin out its product variety. Figure 3.2 shows the major product platforms and model variations for the Walkman

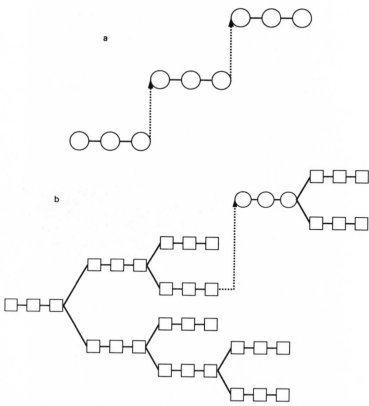

Figure 3.1. Paradigms for management of product families: *(a)* Revolutionary/discontinuous. *(b)* Evolutionary/continuous.

family. On the vertical axis is the average price of Walkman models. On the horizontal axis is the year in which the advertisement appeared (see Sanderson and Uzumeri, 1990, for more complete discussion).

These platforms provide a basic mechanical and electronic core that is altered and enhanced to produce variants with different features or external appearances. Each of these platforms optimizes a particular design goal that is matched to market needs. For example, high quality and small size Walkmans are important in the Japanese market and many of the Walkmans produced for Japan are derived from the WM-20 and Super Walkman. Low cost is more important in the U.S. market and most Walkmans sold in the U.S. are derived from the WM-2 (Sanderson and Uzumeri, 1992).

Teams made up of design and manufacturing engineers worked jointly to develop both new models and the manufacturing systems to produce them. These teams are responsible for the variations and improvements of the five core designs that make up the bulk of Sony's product introductions. Special ''tiger teams''—separate from those executing product improvements and variations—conduct aggressive technological development. The technological teams are given clear stretch goals and longer time frames. Such teams were responsible for the development of the WM-20

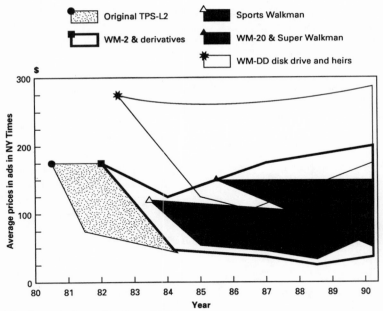

Figure 3.2. Core platforms for Sony Walkman.

model, a Walkman that is only slightly larger than the cassette it plays. Other technological advances in the Walkman include ultralight headphones, super-flat motors, and rechargeable batteries the size and shape of a stick of chewing gum. These core devices have become the basis on which future models will be built.

Sony deploys marketing/industrial design teams in Japan and in major regional markets to develop new product concepts attuned to market needs and trends. These groups develop product specifications and mock-ups, which are then transferred to the engineers in Japan for engineering implementation, via modifications of one of the core device designs. The U.S. team originated niche designs such as the My First Sony line of children's products and the rugged-style sports and Outback series.

The heart of Sony's strategy was one of incremental innovation—a series of rapid product changes of limited scope designed to satisfy different customer needs and keep the product line fresh. At the same time, significant technical advances (the super-miniature Walkman, the rechargeable Walkman, and the direct-drive Walkman) and an ability to spot and exploit different patterns of customer preference contributed to Sony's success in the Walkman family. In addition to an evolutionary pattern of product development, Sony has set clear goals and pursued specific design challenges in the development of its product families.

Setting Clear Goals

Fundamental to effective approaches to design for manufacturing in an environment of continuous improvement and change is a clear sense of direction. Continuous product improvement must have both a basic *concept* and a *direction* for change. Sony has developed a strategy that guides the development of many of its consumer

Table 3.1. Discman© Design Concepts and Direction

Concept	Direction
Anyone	Smaller
Anytime	Save power consumption
Anywhere	Light
	Digital

products. A key element in that strategy has been to take the new products and extend applications by developing highly portable products that can be enjoyed, according to the Sony credo, "anytime, anywhere, and by anyone." Once the overall direction of innovation has been established, Sony sets difficult specific goals and challenges its engineers to meet them. Sony's product families have been designed to follow a trajectory toward smaller, lighter, digital products that save on the consumption of power.

Japanese firms in consumer electronics markets have generally had a clearer sense of direction than American firms because of the strong cultural preference for small compact products of high quality with enhanced features. Japanese consumers are, on the whole, less price sensitive that American consumers. They turn products over more quickly and are willing to pay for enhanced features and superior performance. As a result, design engineers in Japan have tended to have a better and more consistent sense of what products and features Japanese consumers will find attractive.

Pursuing a Specific Challenge: The Direction of Change

Sony's portable CD player was introduced in 1984. Introduced only two years after the initial appearance of the compact disc, the Discman, the world's smallest CD player, has become the universal symbol of the portable CD player. At the time of its appearance, the only other CD players available were normal-sized components, and the majority of customers were audiophiles at the high end of the market. A key achievement of the portable CD was reliable sensing and control of critical optical and mechanical components. Sony's D-50, featuring a refined version of the high-density chip-mounting technology first used in Sony's Walkman, opened the compact disc market to a much wider audience. The Discman was designed to be enjoyed "anytime, anywhere, and by anyone." Table 3.1 illustrates the concept and direction of change for the portable CD player.

The original Discman (D-50) was 127 (W) by 36.9 (H) by 132.5 (D) mm, used 4 watts of power, and weighed approximately 590 grams (without batteries). In 1985 Sony introduced the D-50MkII with a body volume reduced by 40% and weight reduced by 14%. Its central feature was a reduction in power consumption to 2.6 # watts, some 35% less than the D-50 required, allowing it to be used outdoors with eight size AA (LR6) batteries. In addition, this model opened the way for true multifunctionality in a portable CD player, featuring a repeat-play mode and 16 RMS (random music sensor) for the first time. The drive to make the Discman portable

Table 3.2. Sony's Major Portable CD Models

Type	Deck	Portable	Pocket size
Model	CDP-101	D-50	D-88
Introduced in Japan	October 1982	November 1984	April 1988
Dimensions (mm)	355 x 105	127 x 36.9	94.5 x 32.9
(W × H × D)	x 325	x 132.5	x 99
Volume (cm³)	12,115	585	277
(Ratio)	(1)	(1/20)	(1/44)
Weight (g)	7,600	590	300
(Ratio)	(1)	(1/13)	(1/25)
Power consumption (w)	23	4	1.7
IC	42	29	14
(incl. LSI)	(3)	(4)	(7)
Laser pickup size (Ratio)	(1)	(1/3)	(1/5)
Number of parts	~1,000	~600	~400

CD player even thinner, lighter, and easier to use resulted in the development of the D-100, a player that broke the 20-mm barrier for the first time, with a height of only 19.8-mm and a weight of only 420 grams.

In 1988, Sony introduced the first truly pocket-sized portable CD player. The appearance of the 8-cm CD as a digital replacement for the former 17-cm analog "single" signaled the final transition from analog to digital modes, from records to compact discs. Through the use of high-density chip-mounting technology, miniaturization of printed circuit boards, reduction in the number of parts, and a new no-chassis structure, the D-88 achieved ultra-compact size in an ultralightweight body of only 300 grams. The D-88 was designed as a compatible model to allow the playing of standard 12-cm compact discs as well. Table 3.2 summarizes the body dimensions, weight, power consumption, and number of parts in the successive Discman models.

Sony engineers sought to produce the best-quality sound in the smallest possible space. In designing these Discman models Sony had set for itself a clear direction— to manufacture smaller (power-saving) lighter, digital products. Sony organized its overall engineering design goals to meet this challenge, and successive upgrades reduced power consumption, weight, and overall size (see Figures 3.3 and 3.4).

Much of Sony's design effort has focused on decreasing power consumption, decreasing the overall dimensions, and reducing weight. They have pursued this goal assiduously, making improvements in this direction with each successive series of models. Their ultimate goal was to produce a device only slightly larger than the disc it played.

Producing for all Lifestyles

In addition to producing a highly portable CD player, the present lineup of models also covers a wide variety of prices ranging from 19,500 to 64,000 yen and a spec-

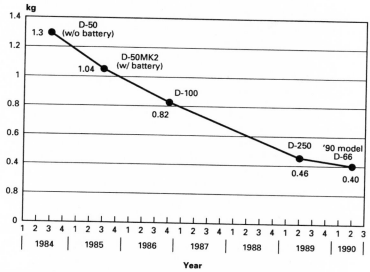

Figure 3.3. Weight of Sony's Portable CD Models.

trum of variations to meet different customer needs and lifestyles (see Table 3.3). Sony designs portable CD players for use in the home, outdoors, the car—anywhere people listen to music.

Product planning plays a central and critical role in the actual selection of features and models to be developed. The product planning group meets two times a year to decide on future directions. It is the product planning group that finally determines which features to promote and what price targets to set. They take the information from the marketing division and salespeople concerning what features

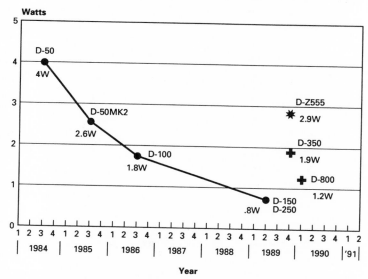

Figure 3.4. Power Consumption in Sony's Portable CD Models.

Table 3.3. Sony Portable CD Models Available
in Japan, 1990

Target	Model	Price (¥)
Indoors	D-Z555	64,000
	D-350	43,000
Outdoors	D-T20	32,800
	D22	19,500
Auto	D-800K	36,000
	D-810	27,500
Everywhere	D90	34,900
	D99	32,800
Pocket	D-82	20,300

people like and what they don't like. Often engineers themselves visit disgruntled customers to try to understand the problems consumers are having with their products.

While most product planning is initiated by the product divisions and sent to electronics, mechanical, and physical design groups and then to production, on occasion product planning at Sony has been initiated by the industrial design center. A new product idea or concept is given to electronics and mechanical design and then to production. The industrial design group has driven design in the development of some models of the Watchman, the Walkman and beta VCR.

A strategic approach to product planning and clear setting of direction and goals for the management of product families is essential for continuous innovation. As product and process cycles shorten there is a need to develop new computer-based techniques as well as organizational forms to permit flexible production of multiple models on the same production facility with low cost and rapid changeover between production cycles. The use of product strategies that take advantage of product families can significantly lower both the costs of product development and manufacture, but they have been underutilized by U.S. firms because they have not yet developed the methodologies for their effective deployment (Nevala, 1989).

From this perspective, the changes associated with new product design should be a planned rather than a reactive process. Although it is now well recognized that early involvement by manufacturing and other key personnel aids in the conceptualization of new products and improvements, most firms are not currently organized to handle the integration of the large amounts of diverse and complex information, which includes not only considerations of product form, function, and fabrication but also the organizational and administrative procedures that underlie the design and engineering process.

U.S. firms have been weaker than their Japanese counterparts in managing the design, manufacturing, and marketing interfaces effectively. They have lacked the discipline to innovate incrementally and get the most out of original technical insights and product ideas. They have been successful in developing initial product prototypes but have generally not had the patience or set clear enough design and manufacturing goals for the effective development of product families. Nor have they been able to achieve the level of cooperation among disparate members of the organization and among independent project teams to focus their efforts on goals that go beyond the scope of individual projects.

What the Japanese have lacked in tools they have made up in good management relations and a disciplined approach to the management of product families. The development of new design tools holds some hope for firms that have not been able to achieve the success that Japanese firms have had in new product development. But they too will have to adopt a disciplined approach to the management of product families if they are to use these tools to their fullest and achieve competitive success.

THE ROLE OF DESIGN TOOLS

Managers and engineers in the most progressive companies in the world have recognized the need to incorporate manufacturability concepts into the design of products. Consumer electronics and computer companies have been very proactive in their efforts. They have understood the power of the computer and software tools to help both design and manufacturing engineers do analysis to review new designs for the purpose of optimizing them for manufacturing. Several computer firms, including IBM and DEC, have supported internal as well as university research on design for assembly and design for manufacturing. They have made major internal efforts in the development of software for modeling cost and technology trade-offs for printed wire assembly as well as other applications.

Despite their best efforts, the original intent to use these early design tools on a broad spectrum of products has been difficult to achieve. Although some techniques have been incorporated into the normal manufacturing review system, there has been a general reluctance to implement these techniques into the existing design process. There is a pressing need to develop a methodology for designing products, developing manufacturing processes, analyzing the cost impacts of alternative designs, and integrating these decisions into the normal work environment.

Computer networks, common databases, graphical interfaces, and distributed control and software are providing the tools to integrate and coordinate product and systems design environments. These new design tools offer the promise for capturing design knowledge and communicating it between the product design engineer and the manufacturing engineer. However, the effective use of these new tools will depend upon their integration into the complex organizational structure of the design-manufacturing enterprise (Adler, 1989a; Beatty and Gordon, 1988; Majchrzak, Nieva, and Newman, 1986).

In the past, engineers put their main emphasis on the development of tools, and it was assumed that people would take readily to them and incorporate them into their daily work. By the same token, management research has generally ignored the limitations of the tools themselves and assumed that any problems that arose were due to reluctance on the part of the users or limitations in the organizational system in which they were deployed. The technology was taken as a given, and organizational methods were suggested for their more effective deployment. More recently, a number of studies have demonstrated that effective use of advanced design and manufacturing technologies depends both on the technical attributes of systems and on organizational and human factors necessary for their successful adoption and use (see, for example, Ettlie and Reifeis, 1987; Majchrzak and Salzman, 1989). Despite this growing recognition, few organizations are willing to make procedural, organizational, and cultural changes to achieve more effective integration of design and manufacturing (Adler, 1989b; Badham, 1989; Majchrzak and Salzman, 1989). Even less attention has been given to ways the technology can be designed better to complement the needs of the users and more adequately address key business issues.

It is generally believed, or at least hoped, that new information tools will reduce the need for face-to-face communication, making it possible to carry out effective new product design that is truly producible in geographically dispersed design and manufacturing facilities. Many major U.S. firms have manufacturing plants all over the world. Moreover, in order to create effective windows on new technology as well as develop a better understanding of customer needs, U.S. firms have set up design centers in Japan and Europe and in countries such as Singapore with an abundance of highly skilled engineers. Maintaining the high degree of coordination required to design and manufacture products in geographically dispersed facilities is particularly challenging in the fast-cycle environment characteristic of many products in the electrical, electronics, and computer industries.

There has been significant development of new design technologies over the last ten years. These technologies have focused on three major areas: database technology, networking and communication technology, and automated design and planning tools.

Database Technology

The technology for design databases has focused on the representation of products, components, and design principles in order to provide a basis for communication of results and for the interactive design of new products. Overall design methodologies have been divided into two major groups: (1) those that focus on electronic system design and (2) those that focus on mechanical system design.

Database issues in both electronics and mechanical design are fundamental to the effective communication of the design process and results. Participation in the design of complex systems involves large teams of individuals, each of whom is designing a small portion of the system. Each designer makes decisions that will locally optimize the design but not necessarily contribute to the overall function, efficiency, or manufacturing ease of the system as a whole.

The emphasis on the development of database systems has been to provide a capability so that when one designer presents his design decision or changes, other designers can easily integrate them. This capability makes it possible for the new overall design to be more easily upgraded by the integration of these changes, so that the results of the design process can be transferred to the next stage in the process. When a schematic designer transfers his results to someone who is designing silicon chips, much of this process is made possible by compatibility of the electronic design database. In the mechanical design world, this transfer is much more difficult since there is no general direct compatibility between the mechanical description of parts and the program that is used to drive the machine tools to manufacture the parts.

Recent efforts in electronics design technology have emphasized the development of so-called silicon compilers, which are integrated design database systems that accept functional descriptions of electronic systems and automatically generate a series of levels of description culminating in a specification for the implementation of the electronic system into a silicon chip. Major questions in electronic system development still remain for the integrated database description of larger electronic systems such as power supplies, circuit boards, and computers. In these systems, the physical configuration of parts and devices is more difficult to describe and interfaces with problems in mechanical design system development.

In mechanical design the database description has been much more problematic. The effective description of mechanical parts in such a way that the parts may be represented on a graphics screen is quite different from the design database problem associated with the description of parts to be manufactured, and different still from the description of parts for automated inspection. The commercial tools for mechanical design are oriented for display of graphical information only. Some of the newer efforts, such as the product description exchange standard (PDES), are oriented toward the development of common informational frameworks that ensure the compatibility of information among different companies' systems design stages and ensure the ability to develop interfaces to process technologies that are uniform across different products.

The PDES and related efforts that have been discussed in the development of concurrent engineering, are examples of new database technology that is only beginning to have its impact on practical industrial uses. It is also clear that the effectiveness of these technical developments will be strongly determined by the organizational and managerial environment in which they are implemented.

Networking and Communication Technologies

The shared database provides an electronic mechanism to store the design as it evolves. It provides a structured electronic format in which to capture design decisions as they evolve. The database can exist in a number of forms and the associated technical developments will continue to require significant research and innovation in order to provide the appropriate sets of representation for design and the appropriate interfaces for different members of the design team.

Communication takes place by access to common data storage. One may utilize this mechanism either with independent organizations or joint working groups. The question that is often addressed is whether the shared database mechanism is sufficient to provide communication between independent organizations or whether the personal interaction of the working groups is necessary to provide the required communication. The basic principle of the shared database is that voluntary access provides an ongoing view of the evolutionary design decisions and therefore will provide an opportunity for working groups and different members of the design team to provide their own input as well as to anticipate the consequences of those design decisions.

Networking may further enhance these objectives (see Figure 3.5), particularly if combined with a shared database. The networking mechanism provides an electronic means to capture this participation and anticipation function. The design team members receive electronic messages that describe not only the design decisions but

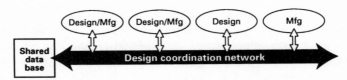

Figure 3.5. Tiger teams or functional groups sharing a single database and using design coordination network.

the rationale for design decisions. Thus, when accessing the structured shared database, they have network access also to the rationale behind design decisions.

In Japan, Sony has developed a direct electronic link and shared database between its industrial designers and suppliers of the plastic and metal casings used in its audio and video products. In this way Sony has been able to reduce the time it takes to prototype and develop new designs for its fast-cycle products such as the Walkman.

In a sophisticated model of these systems, such a network would also communicate the implications of design decisions selectively to different members of the team. This enhanced networking capability could be envisioned to provide a level of communication that would enhance the interaction among the team members in both of the management dimensions we have described.

There is significant technological development needed in order to provide this combination of shared database and networking tools. A number of unanswered questions persist: To what extent would such a system be effective if used by an organization with design team members located outside of the same complex or in different parts of the world? Is it necessary to bring working teams together in order to provide the proper kind of communication?

Shared databases and communications technologies are the major new tools that are becoming available and have the potential to change the ways in which design and manufacturing engineers have interacted in the past. A number of important questions need to be addressed before they can be effectively deployed. How should they be incorporated into the design and manufacturing organizations? What are the optimal organizational forms for their use? Who should have access to the network or the shared database? Who should have control and responsibility for maintaining and updating the systems? What role should design rules play in the design process? Who should have responsibility for defining design rules and how should they be updated and maintained?

Automated Design and Planning Tools

Planning and reasoning systems automate design decisions based on the database description of the parts and products, as well as rules that are implemented to make design decisions. For electronic systems such rules may be as simple as those that ensure the connection of wires between points or may be as complicated as sophisticated models of the dynamics of electrons in individual devices. In mechanical design such rules often involve the detection of overlap between objects and assemblies, the specification of tolerances among parts, or the order in which machining operations can be run in order to manufacture a part that has been physically described.

Such planning and reasoning tools are relatively advanced in electronics, where the specification of relationships in the electronic circuitry are much more clearly defined or can be specified or constrained under certain manufacturing processes to follow very carefully.

It is common practice in electronic design to require the designer to follow specifications so that the design rules will be consistent and implementable. This careful constraint over the design process is not as common in mechanical design; and the freedom of the designer to choose parts, specifications, tolerances, and so on, leads to a much larger variety in the types of designs that evolve. This latitude has been one of the major criticisms of the efficiencies of the design-to-manufacture step. Despite its limitations, automated design and planning tools have been used successfully in mechanical design to simulate manufacturing processes to predict the impact of product design decisions in the development of body panels at the Toyota Motor Corporation.

The design of the body and its production preparation are critical in determining lead time in the development of a new car. Much of the lead time and engineering manpower are involved in the design and manufacture of the stamping dies. By developing a computer-aided engineering (CAE) system for stamping dies, the Toyota Motor Corporation has been able to achieve significant time and cost improvements in the simulation of body panels.

Prior to the development of the CAE system, clay models of the idea sketches of die designers were prepared and converted to life-sized exterior drawings, which were then sent to the body design group. Information on the body structure and complex free surfaces were put into the life-sized master body drawing and the production preparation department would read out the body drawing with complicated free curves. Master models were then prepared for individual parts, and a similar information-reading process was repeated in the design of stamping dies. The rough geometry of the stamping die surface was then drawn around the lines, the clay was applied around each master model according to the rough drawing, and the designer would determine the die face shape by checking on necessary items. Plaster reproductions were used as reference models in the subsequent machining, finishing, and inspection of the stamping dies. A number of tryouts and die face modifications would have to be repeated against forming defects and geometric inaccuracy in order to create final stamping dies suitable for mass production (Takahashi and Okamoto, n.d.).

Many engineering man-hours and much effort were spent in the transmission and processing of such information. Toyota engineers developed and introduced computer-aided design/manufacturing (CAD/CAM) systems and a numerical control (NC) processing system for stamping dies and, finally, a CAD system for styling and body structure design. Subsequently a CAE system was developed to simulate press forming defects in body panel designs with a resulting significant reduction in lead times and engineering manpower. The first practical application of the stamping CAE system took place in 1981.

According to Toyota engineers (Takahashi and Okamoto, n.d.), die design speed has been accelerated by the evaluation of press forming severity through simulation,

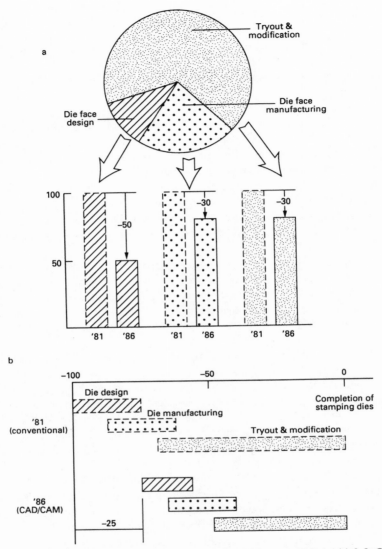

Figure 3.6. *(a)* Man-hour reduction by "die face CAD." *Source* A. Takahashi & I. Okamoto, Toyoto Motor Corporation. *(b)* Lead time reduction by "die face CAD."

resulting in a 50% reduction in man-hours (see Figure 3.6a). The elimination of measuring work required for the fabrication of models and the preparation of NC processing data in the manufacture of die faces reduced engineering man-hours by 30%. In tryout and modification work, engineering man-hours were reduced by 30% by upgrading the evaluation function of press-forming severity, accumulation of technical knowhow, improvement in design accuracy, and upgrading the press die accuracy. The total CAD/CAM system for the body development has led to a lead-time reduction for die design and manufacture of 25% (see Figure 3.6b).

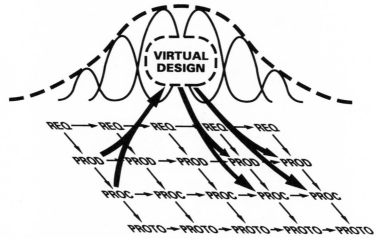

Figure 3.7. Multicycle concurrent engineering.

USE OF DESIGN TOOLS TO MANAGE CONTINUOUS OR EVOLUTIONARY DESIGN

While the design tools described above can be effective in managing either revolutionary or evolutionary design paradigms, some special approaches may be particularly useful in managing evolutionary or continuous design. Multicycle concurrent engineering and virtual design are two such approaches that have the potential for more effective management of evolutionary design.

Multicycle Concurrent Engineering

Managing consecutive projects such as those that might be used in the evolution and development of a family of products is a major challenge. Because the focus has been on the management of independent projects, most scholars have underestimated the role of design discipline, design rules and tools to incrementally iterating a product family. What is needed is an approach to managing product families over the life cycle of the product and to carry out consecutive engineering projects effectively. Such an approach is illustrated in Figure 3.7.

Figure 3.7 shows a model of concurrent engineering in which the design of successive models must be managed over time. In single-cycle concurrent engineering (a one-model cycle) there is a phased overlapping of the design activities. However, managing a product family in which new models are produced frequently (for example, every year) over a period of years, presents a significant additional challenge. Requirements for a new model must be specified at the same time that procedures and manufacturing tooling are being specified in the model currently being designed. The diagram proposes the concept of ''virtual design'' (described below) as one means for easing the transition between cycles. Holding over information at a high level of abstraction about previous designs may prove to be an effective way to make a rapid and smooth transition between product generations. This concept is described in some detail in Sanderson (1990).

But holding over information can be difficult and costly. Most firms are not currently organized to handle the integration of the large amounts of diverse and complex information, which includes not only considerations of product form, function, and fabrication but also the organizational and administrative procedures that underlie the design and engineering processes.

The successful introduction of multicycle concurrent design into organizations requires an investment in new design tools and the integration of productivity, reliability, and diagnostics into CAD/CAE-supported design processes, CAM, as well as new organizational techniques and methods. One common feature of the development of these tools is the search for a consistent representation of the design elements in a way that will allow the current design of a system to be carried over into the design of related products and systems (virtual design). The emphasis on the reuse of design representations and the carryover of common design elements is intended to maximize the efficiency of the designers' time so that they don't have to replicate the design of parts of a system that have already been shown to work correctly. It also carries over to the manufacturing process and permits the manufacturing system to carry over common elements that have been successful in the past.

Virtual Design

New capabilities in information processing and computer-aided design technologies are making it possible for firms to store information about product design and function and to reuse that information in designing new products for model changeover. These design and communication tools have the potential to be utilized as the foundation for a virtual design process. In previous work, I have described the concepts of virtual design and the virtual product cycle as a generalization of a class of emerging design tools and as a conceptual framework for the development and evaluation of new methodologies for the design and manufacturing process (see Sanderson, 1989). Specifically, virtual design is a process in which all of the information about the product and its manufacture is captured in a virtual (computer) environment where evaluations and changes can be easily handled.

New computer-based technologies such as shared databases and networking permit design representations that span several cycles of product designs. Virtual design exists as a computer representation of function and physical description. The virtual product cycle describes the cycle of automated product design representation, which permits flexible transmission and evolution of designs. Similarly, the virtual process cycle describes the cycle of automated process control that is involved to manufacture products with a particular set of manufacturing resources. Again, computer-based technologies generate descriptions for a family of processes.

In virtual design models, off-line computer representations of product designs and manufacturing processes evolve with the product, permitting a graceful transition between design and the integration of process technologies. The resulting virtual product cycle is more gradual and amenable to organizational and strategic planning. Such a virtual product cycle is illustrated schematically in Figure 3.8, where the dashed line suggests a virtual product cycle that spans the rapid changes in product design.

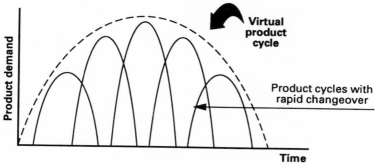

Figure 3.8. Rapid and virtual product development cycles.

Although virtual design concepts have only just begun to influence firms, examples of virtual design can be found in the computer software industry, where program libraries and routines developed for one program are often modified and used in other programs and in the semiconductor industry in gate arrays and applications-specific integrated circuits (ASICs). Program libraries are already in use in the computer and semiconductor industries and will expand to other industries as CAD systems take root and become an integral part of the design and manufacturing process.

It is more difficult to find examples of virtual products in electromechanical applications since sophisticated CAD technologies are not yet widely used in product design, and few people have tried to rationalize the design process. More typically, as we saw in the case of Sony's Walkman family, firms use standardized parts and components across models and distinct products (group technology) in an effort to make manufacturing simpler and less costly. Sony used product development platforms and spun out a wide variety of products targeted to particular market niches effectively using standard modules combined with proprietary building blocks—generally sophisticated, specialized electronics and circuit boards.

Using standard modules with an appropriate mix of revised and new parts, companies can develop products with unique characteristics of various models quickly. Gomory (1989) reported that one group at IBM, working on a series of display terminals, trimmed five months off the previous development cycle for the same line of products by just using standard components. Sony routinely uses its core devices to build product variety and features targeted to different market segments.

Virtual design is a step beyond group technology and standard modules and components, as it allows for the modification of designs in real time. By incorporating the virtual design process into the design and manufacturing cycle, a firm may be able to improve the efficiency of both the design and manufacturing implementation processes and gain economically through a more timely response to market changes and the need to meet a variety of customer needs.

CONCLUSIONS

U.S. firms increasingly find that their methods of product development, from design to manufacturing to marketing, cannot react quickly enough nor accommodate the diversity necessary to satisfy the demands of international competition. Making incremental changes and getting a large number of models into the market quickly and in a cost effective manner is the only way firms in these industries can retain market share and generate profits (see Sanderson and Uzumeri, 1990). Upgrades and extensions of initial product designs and concepts may be the keys to success in highly competitive environments like consumer electronics and computers.

Organizational solutions to the problems of DFM tend to focus on the formation of special design and manufacturing teams with clear goals to design a producible product. Once this goal is achieved and a prototype is successfully designed and manufactured, the teams often disband and the firm goes back to business as usual. Upgrades and redesigns frequently do not command the same attention or resources as the initial development, and as the product evolves, upgrades may not be as producible as the initial design. Few American firms have developed formal mechanisms to assure DFM for upgrades and extensions of existing products—even ones that were initially designed using DFM principles. Moreover, there is a tendency among American engineers to seek their own unique solution to technical problems that makes sharing among projects difficult.

Truly successful new product design and product renewal requires the development of the organizational and technical capability of carrying out DFM all the time, in initial designs as well as in upgrades and renewals of the original products. American and European firms that have found it difficult to match Japanese-style relationships (both within and between organizations) are looking to emerging design and manufacturing technologies to improve integration of these functions. If current and emerging tools are developed and implemented with more management insight than some of their predecessors, it is likely that firms will be able to achieve significant time, cost, and quality improvements and may even be able to match the Japanese in the effectiveness of their design and manufacturing systems. It is clear, however, that the development of these paradigms depends on the effective use of new tools that, in turn, depend as much on their management and integration into the organization as on their intrinsic technical merit.

The process of evolutionary or continuous innovation, with the aid of design tools, can help significantly to achieve design for manufacturing in an environment of continuous change such as that which characterizes many of today's fast-cycle products. But it is only through the effective management of product families that firms can hope to create a dominant position in the market over any sustained period of time. Firms that invest in the capacity to make changes and ease the process of transition for one product generation to the next can create a major competitive advantage.

REFERENCES

Adler, Paul S. "CAD/CAM Integration and the Design/Manufacturing Interface." Stanford University, Unpublished Working Paper, August 1989a.

———. "CAD/CAM: Managerial Challenges and Research Issues." *IEEE Transactions on Engineering Management* 36(2) August 1989b, 202–215.

Badham, Richard. "Computer-Aided Design, Work Organization, and the Integrated Factory." *IEEE Transactions on Engineering Management*, August 1989, 216–226.

Beatty, Carol A., and John R. M. Gordon. "Barriers to the Implementation of CAD/CAM Systems." *Sloan Management Review* 29(4), Summer 1988, 25–33.

Ettlie, John E., and Stacy A. Reifeis. "Integrating Design and Manufacturing to Deploy Advanced Manufacturing Technology." *Interfaces* 17(6), November-December 1987, 63–74.

Gomory, Ralph. "From the 'Ladder of Science' to the Product Development Cycle." *Harvard Business Review* 67(6), November-December 1989, 99–105.

Majchrzak, Ann, Veronica F. Nieva, and Paul D. Newman. "Adoption and Use of Computerized Manufacturing Technology: A National Survey." In *Managing Technological Innovation*, ed. Donald D. Davis. San Francisco: Jossey-Bass, 1986, 105–126.

——— and Harold Salzman. "Introduction to the Special Issue: Social and Organizational Dimensions of Computer-Aided Design." *IEEE Transactions on Engineering Management*, August 1989, 174–179.

Nevala, David. "Design for Manufacturing and it's Relationship to Concurrent Engineering." Presented at Computer Integrated Manufacturing Annual Meeting, Rensselaer Polytechnic Institute, November 10, 1989.

Sanderson, Susan W. "Strategies for New Product Design and Product Renewal." *Academy of Management Best Papers Proceedings*, 1989, 296–300.

———. "Time and Cost Models for Managing Change and Variety in a Fast Product Cycle Environment." Center for Science and Technology Policy, Rensselaer Polytechnic Institute, 1990.

——— and Vic Uzumeri. "Strategies for New Product Development and Renewal, Design-Based Incrementalism." Center for Science and Technology Policy, Rensselaer Polytechnic Institute, 1990.

Sanderson, S. W., and Vic Uzumeri. "Industrial Design: The Leading Edge of Product Development for World Markets." *Design Management Journal*, 3(2), 1992, 28–34.

Shintaku, Junjiro. "Technological Innovation and Product Evolution: Theoretical Model and its Applications." *Gakushuin Economic Papers* 26(3), January 4, 1990.

Takahashi, A., and I. Okamoto. "Computer-Aided Engineering in Body Stamping." Toyota Motor Corporation, Japan, n.d.

Womack, James, Daniel Jones, and Daniel Roos. *The Machine That Changed the World.* New York: Rawson Associates, 1990, 119, 121.

4

PRODUCTIVITY IN THE PROCESS
OF PRODUCT DEVELOPMENT—
AN ENGINEERING PERSPECTIVE

PHILIP BARKAN

According to many recent assessments, the best U.S. industries are now approaching parity with Japanese competitors in manufacturing costs, manufacturing defect rates, and product quality. The picture is less sanguine in terms of engineering productivity as measured by the total elapsed time required to develop new products, by the total engineering man-hours required to develop new products, and by the quality of the initial product design as measured by design changes late in the cycle, field defects, and recalls.

On these scores, U.S. manufacturers generally remain deficient with respect to their Japanese counterparts. The persistence of such problems raises questions about U.S. engineering productivity and U.S. engineering effectiveness. Why should U.S. engineers require up to twice the engineering manpower for twice the time in order to design comparable products? In this chapter possible reasons for this discrepancy will be examined, ultimately focusing on engineering changes as both a symptom and cause of the basic problem. Possible strategies for attacking the problem will then be examined.

REASONS FOR THE DISCREPANCY IN
ENGINEERING PRODUCTIVITY

Use of Outside Suppliers

Japanese manufacturers are far more likely to farm out the engineering and production of major components to their satellite suppliers. Clark, Chew, and Fujimoto (1987) indicate that Japanese manufacturers use not quite half the number of engineering hours to produce an automobile of comparable size, style, and price as do American manufacturers. Satellite suppliers account for over 50% of Japanese engineering hours versus only 10% of American engineering hours. However, the Japanese advantage remains even after creating hours to account for the role of suppliers.

Longer Working Hours

Japanese engineers also routinely spend much longer hours on the job. While the occasional 80- to 100-hour week is not unknown to U.S. engineers, the effectiveness of a sustained schedule of long hours can be questioned. However, some possible side benefits can be derived from such a schedule such as close camaraderie and improved communication.

Overlapping Activities

Clark, Chew, and Fujimoto (see Chapter 11), in their worldwide study of the automobile industry, found that important differences in engineering productivity derive from a greater use of overlapping activities. In Japan, long lead-time tooling is released and rough machining initiated well before the parts are fully designed. But such practices are uncommon in many Western industries, where the benefits of reduced time to market are not perceived as ample justification for the risk of financial loss from the scrapping or modification of tools.

Simplified Procedures

Japanese manufacturers operate with far fewer layers of administration, require far less paperwork, and accrue a significant benefit in productivity and time by major departures from our accepted procedures. Here is how Hammer (1990) describes the astonishment of the Ford Motor Company when it benchmarked its accounts-payable operation against the comparable Mazda Motor Co. operation:

> The difference in absolute numbers was astounding, and even after adjusting for its smaller size Ford figured its accounts payable operation was five times the size it should be. The Ford team knew better than to attribute the discrepancy to calisthenics, company songs or low interest rates. . . . Where it has instituted the new process Ford has achieved a 75% reduction in head count.

Hammer emphasizes the need to "use the power of modern information technology to radically redesign our business processes in order to achieve dramatic improvements in performance." If analogy is any guide, we can expect to find similar opportunities in engineering as well.

Following this guideline, Japanese companies build on their close relationship with suppliers to eliminate much of the time consumed in non-value-added activities such as

- tedious procedures of competitive bids
- arcane negotiation of complex, formal contracts
- laborious processing of purchase orders
- routine inspection of goods from qualified suppliers

There exist within every engineering organization activities whose value can similarly be questioned.

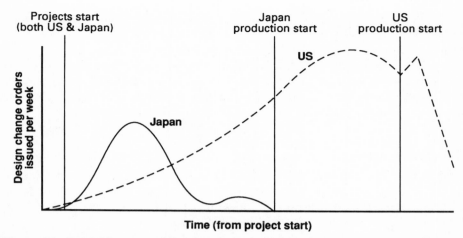

Figure 4.1. Rate of issuance of design changes—patterns of U.S. and Japanese auto manufacturers. Source: Sullivan (1987).

Better Control of Engineering Design Changes

The need for engineering design changes is inherent in the development process, both in Japan and the United States. Striking differences in the quantity, quality, and timing of design changes, however, account for a significant proportion of the U.S. shortfall in productivity. The role of engineering changes as an important cause of the productivity discrepancy is identified in studies by Sullivan (1987) of the auto industry and by Bebb (1990) of the photocopier industry.

According to Bebb's study of the photocopier industry, the U.S. development process required up to four times as many design changes as well as a longer development cycle. Similar trends, reported by Sullivan in his study of the automotive industry, are graphically illustrated in Figure 4.1. Striking differences are shown in the rate of issuance of engineering changes (i.e., changes issued per week) as a function of time in the product development cycle.

Note in Figure 4.1 that Japanese automobile companies process many more changes per week in the early stage of product development. The rate reaches its peak roughly one-third of the way through the total development period, then rapidly decays to minor proportions about two-thirds of the way through the development cycle.

In U.S. companies, the issuance of change orders peaks very late—about 90% of the way through the development period, as contrasted with a Japanese peak at about 35%. Changes continue to be issued at a high rate even as the product is released and well into the production period.

The differences in quantity and timing of design changes disclosed in this curve accounts for much of our productivity problem. However, the significance of this chart goes well beyond the simple count of the total number or rate of issuance of engineering design change notices. Most important, these curves are revealing of another major dimension of the problem—the *character or scope* of the changes.

Early changes are easy to accommodate—their "domino effect" on the total system is relatively minor because interrelationships have not yet been rigidly defined and the vast majority of tooling either has not been released or is still being rough cut. Changes issued late in the product development period, on the other hand, have profound repercussions affecting both the product and the tooling.

Optimum design balance can only be achieved in the early conceptual stage of a design, so that the impact of late change ranges far beyond details and can affect the essential character of the product. When the rate of issuance of change orders is low as the project approaches product release time, it strongly implies that these changes relate to fine tuning of isolated details. On the other hand, massive change implies that extensive, possibly basic modification is going on at a furious rate, often under chaotic circumstances, with great losses in engineering efficiency, deteriorating quality of the product, and drastically increased cost.

WHAT ACCOUNTS FOR THE DIFFERENCE IN PATTERNS OF ENGINEERING CHANGE NOTICES?

While the significance of other factors should not be dismissed, engineering changes are without doubt a primary problem. For this reason the remainder of this chapter is devoted to an examination of the problem of engineering changes, their sources, and possible means of control. It is proposed that the pattern and character of engineering changes is an important measure of the effectiveness of the entire project operation—ranging from its initial inception to its detailed implementation. That is, the sources of engineering changes extend well beyond the engineering function, and reflect shortfalls in the planning, staffing and execution of virtually all segments of the project.

If distinctions in engineering change notices are so important a factor in accounting for the shortfall in engineering design productivity, the key question then becomes, What factors are responsible for the vastly different quantity, timing, and character of design changes? The following factors are proposed to account for the differences in patterns of design changes.

Completeness of the Product Definition

One of the most important issues to be resolved at the outset of a project is a sound definition of the product in a broad statement that captures the character of the product as well as the essence of the distinctive features needed to assure its success. There is a wide consensus that one of the greatest sources of weakness in projects follows from the lack of an adequate project definition.

Wilson (1990) in an extensive study of Hewlett Packard projects differentiated successful from unsuccessful programs solely on the basis of consistency in satisfying ten factors related to product definition:

1. The strategic intent of the product is refined.
2. The customer base and customer wants are identified.
3. Priorities that facilitate decision making are set.
4. Regulatory issues are addressed.
5. Competitive benchmarking is carried out.

6. Competitive moves are anticipated.
7. Common understanding of project goals is addressed.
8. The right technology is utilized.
9. The best manufacturing/production processes are selected.
10. Broad organizational endorsement and support are gained.

Problems arising from changes in product definition are probably more likely when following reactive strategies, that is, a product design whose features are driven by competitors' strengths as contrasted to proactive strategies, which are driven by internal strengths. In reactive responses there is greater pressure to incorporate unproven technology that has not been fully refined and which therefore entails the risk of causing basic problems downstream in the development process, requiring further change and further degrading product quality and engineering efficiency.

Restriction of Design Innovations to Proven Technology

To clarify the intent of this discussion, it is first necessary to distinguish between "research and development" and "product engineering." These are distinct and different activities—both are essential to the enterprise.

Research and development deliberately seeks to confront the frontiers of knowledge and corporate experience. It seeks to create new knowledge and relative certainty which makes possible subsequent application of this knowledge to produce high quality, reliable products. In R&D we expect and accept a high degree of uncertainty, error, and failure.

Product engineering seeks to create with a high degree of certainty and in minimum time superior products measured in terms of cost, quality, feature, and responsiveness to customer needs. Minimum time and highest efficiency are only possible if we avoid failure and seek to minimize uncertainty and risk by the disciplined use of proven methodologies and proven technology.

Many massive design changes arise from the temptation to "hit a home run," to leapfrog competition by implementing a new technology before it has been fully proven. Particularly as U.S. producers find themselves pressed by competition, this looms as an attractive response—and in many cases may be a sound strategy. But it cannot be safely combined with an emphasis on time compression and productivity enhancement.

Where major innovations, either in kind or in number, are contemplated, risks should be minimized by combining an appropriately restricted designation of the initial market with a flexible development timetable. The key focus in such cases should not be productivity and time compression, but rather product quality. A new product rushed prematurely to highly competitive world markets may never recover from bad first experiences. A growing recognition of the importance of this distinction is reflected in the conservatism exhibited in recent remarks by several U.S. leaders. For example, Jack Welch of General Electric talks about a greater focus on "walks, bunts and singles rather than home runs" (Tichy and Charan, 1989).

Lewis C. Veraldi of Ford (unpublished document) requires that new technology be implemented into product design only after it has been judged to add true value to product or process, to be essential to world-class *product* status, i.e., not developed as an end in itself to satisfy the egos of designers/inventors, and to be simple enough to be used by the people who will be employing it—designer, manufacturer, or customer.

Fox (1989) and Bebb (1990) of Xerox call for "no surprises," and require that new technology meet the following criteria before it can be implemented in a new product development:

- Thoroughly proven over the required range of design constraints in the required environment
- Meets all performance objectives
- Meets cost projections
- Meets life-cycle requirements—maintenance and service
- Manufacturing people comfortable with the design

Integration of Experience

Most organizations are substantially less than the sum of their parts, owing to failure to reduce experience into a retrievable form or to preserve and transfer information to those within the organization who could derive benefit from it. Loss of such crucial information and experience constitutes a major source of lost productivity in the product development process. It has been estimated that the cost of information presently constitutes over 40% of product cost, and this figure can be expected to grow in the future as we enter the information age (see, for example, Bauman, 1990).

The problem of preserving experience is partly technology and partly a question of attitude, priorities, and management. For example, the effectiveness of technology transfer from an R&D group to a product design group, or from a product design group into production is often impaired by cultural differences, organizational barriers, and attitudes that isolate production-oriented from R&D personnel. As one means for overcoming organizational barriers that impede technology transfer, the Japanese have been inclined to transfer people from R&D into product design and manufacturing rather than attempt only to transmit product information. They are also more disposed to keep the team intact until smooth production is achieved (see, for example, Westney and Sakakibara, 1986).

There are potentially many important lessons to be learned from past product development experience—but the lessons are often forgotten shortly after the end of the project. Recent pioneering research at Xerox (Brown, 1991) emphasizes the informal character of organizational learning and the very difficult problems related to its effective dissemination and means to overcome the innate tendency of organizations to ignore or suppress it.

A number of factors may inhibit engineers from learning from experience. One factor stems from the reluctance of the development teams to review difficulties that in hindsight appear avoidable and therefore seem to reflect unfavorably upon the individuals involved. The problem is not easy to resolve but warrants more attention. Comprehensive reviews carried out in a positive, constructive spirit by knowledgeable people impartial to the outcome may be one way to identify shortfalls and to help formulate improved procedures. Another useful idea is one practiced at Toshiba, where engineers are required to maintain a "problem log," in which important problems are recorded and solutions noted. Such data could be stored in a central computer database and recalled by a key word directory.

A second factor is the lack of an effective system to preserve and propagate experience through the organization or even to take the time to record experience. Because the pressures to move on are always great and the resources limited, it is difficult to find adequate time to carry out analyses and preserve data in retrievable form for a meaningful, constructive postmortem. Because it is difficult to devise a means to preserve and communicate these lessons, much experience is inevitably lost—unless organizations are unusually stable. Relying on individual recall is a short term and local benefit only. Ultimately a computer based system may offer some aid, provided that the resources are applied to develop and maintain such a system.

A third factor is the reluctance of engineers to challenge authority. Jack Welch of General Electric has spoken out strongly in favor of a "shake-out" principle, in which leaders are deliberately and frequently confronted by their key people with frank, critical evaluations of the system and their performance. Specifically, this system encourages questions like Why do you require me to do these wasteful things? (see Tichy and Charan, 1990). Such encounters pose problems that not all managers care to confront and few subordinates wish to risk.

Role of Prototypes in Enhancing Design Productivity

Interesting questions about the role of prototypes in achieving superior design productivity are raised by the different perspectives described by Clark, Chew, and Fujimoto in their analysis of the automotive industry (Chapter 11) and by Bebb (1990) in his analysis of the photocopier industry: Clark and colleagues suggest that Japanese automakers emphasize early problem detection by using prototypes that are characterized by "reasonably high representativeness, not perfection. . . . The rapid construction of many prototypes affords more opportunities to quickly identify and remedy problems."

On the other hand, Bebb (1990) states that in Xerox's studies of the photocopier industry, it was found that "Japanese development teams will typically design, build, test and refine a single set of a prototype unit—and then proceed to the first manufacturing block. A comparable American team may require, say, three full prototype, design, build, test, and retrofit cycles with a total of four times the number of engineering changes spread over a longer development cycle."

The seeming contradiction between these two reports arises from very different interpretations of the concept of prototypes by these two investigators. Clark, Chew, and Fujimoto (1987) are including in their prototype count many partial models that simply test a part or a sub-system for operational feasibility and manufacturability. To Bebb, the prototype is the full system assembly that combines all of the features of the final product. He does not regard rough models for feasibility testing as prototypes. The important point is that in both cases the goal is the rapid realization of a highly functional full working model.

In both cases the Japanese advantage appears to involve a more massive upfront effort employing a superior combination of product definition, innovation focused on a limited number of key features, greater use of experience and better use

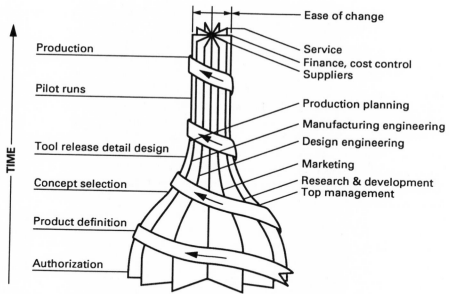

Figure 4.2. The helical model of multi-functional product development.

of analysis, computer modeling, focused innovation, greater integration of manufacturing issues from the start, and quick testing.

Interdisciplinary Experience and Simultaneous Engineering

It is widely accepted today that important benefits can accrue from the formation of multifunctional teams with an intimate sense of project ownership, strong project leadership empowered to make key decisions, and sustained team involvement throughout the life of the project.

Figure 4.2 illustrates the progressive evolution of the design project through its several phases, emphasizing the repeated need for interaction and response by diverse disciplines at every phase.

As illustrated in Figure 4.2, the concept of concurrent engineering requires repeated interaction with all of the disciplines at every phase of the project. The problem is to establish methods for the efficient operation of such cross-functional teams and devise means for enhanced communication of essential information in ways that maximize the efficiency and productivity of the team. Because of the cultural differences between members of multifunctional teams, subtle but significant differences in interpretation can lead to difficulties in execution. Successful implementation of a multifunctional team effort requires the sharing of a set of common values and a consistent interpretation of the project goals. Creating an effective method for imparting this sense widely is an important challenge in team organization and leadership.

While meetings are the commonly used means to improve communication and team effectiveness, these meetings can be a time sink. Finding effective means for communication that reduce the time consumed in meetings has to be regarded as one of the great opportunities for improving engineering productivity. Among the concepts that have proven useful are the following:

- Multifunctional teams that report directly to a capable project leader, empowered to lead and to make key decisions
- Smaller teams with greater scope of responsibility for each team member
- Decision-making and responsibility driven to the lowest practical level
- Team members co-located to facilitate informal communication
- Team members trained in more than one discipline to enhance tacit understanding
- Deliberate controls imposed on the duration and frequency of meetings, and attendance limited to those with direct involvement in the issue
- A common vision imparted by coining a concise statement that captures the essential character of the project

Adler (personal communication, 1988) has suggested that because of their grounding in poetry, specifically haiku, Japanese workers are better able to interpret an abstract vision in meaningful terms.

Structured Methodologies and Disciplined Procedures

Structured methodologies are systematic procedures for fostering comprehensive and orderly thinking on many multidisciplinary design issues. Their use has proven extremely effective when carried out with motivation and care. They help to unify within the design many interrelated issues pertaining to marketing, strategic planning, manufacturing, and quality.

For example, Hauser and Clausing (1988) as well as Sullivan (1987) credit disciplined use of quality function deployment (QFD) as a major factor in the reduction of the need for engineering changes. Major benefits have been derived also from several other structured methodologies, many of which are now gaining serious attention within U.S. companies.

- Quality function deployment
- Robust product and process design
- Value analysis and value engineering
- Tolerance and parameter design optimization
- Design for assembly
- Design for process
- Statistical quality control
- *Poka-yoke* (''foolproof inspection'')

Superior Support Infrastructure

More rapid engineering response and the early detection of problems are keenly dependent on rapid turnaround response from the support infrastructure—including purchasing, model and prototype construction, testing, computer analyses. Further, there is the occasional need for a substantial and timely infusion of additional resources when bottlenecks are encountered. By way of one small example, a comparison by

Clark, Chew, and Fujimoto of U.S. and Japanese car manufacturers reveals that U.S. prototype design extends over a much longer time (8.3 vs. 3.6 months on average) and that prototypes take longer to build (3.4 months vs. 2.6 months on average).

There is reason to believe that similar advantages relate to the use of computer modeling and simulation, as evidenced by the far greater use of high-speed super computers by Japanese industry.

Reduction of Dimensional Errors

Dimensional errors and problems of fit and tolerance are a major source of design problems necessitating change. While computer programs for checking tolerance stack-up are now available, they cannot catch many common errors. For complex geometries there still does not exist CAD software that can catch errors of fit and compatibility adequately and automatically. As a result, many of these errors are only caught late in the design cycle when system prototypes are constructed—often after costly tools have been built.

The ultimate responsibility still devolves upon the capabilities of designers, their willingness to exercise the care and effort involved in rather tedious checking procedures, and their command of such tools as geometric tolerancing and Taguchi's tolerance and parameter design procedures (see, for example, Ross, 1988) for achieving robust designs.

Photolithography is an exciting procedure that can help minimize this problem. It permits rapid creation of solid models directly from the computer. Where applicable, this procedure makes it possible to make plastic models of parts whose compatibility in fit and appearance can be quickly checked.

Setting of well-defined interface boundaries that appear on the drawings of all mating or interacting parts also can eliminate many design compatibility problems by early definition in writing of the quantitative requirements at the interface boundaries. Properly executed, such written definition of boundaries can obviate the need for repeated meetings and eliminate the need for changes brought on by interface problems. This idea exemplifies the *poka-yoke* or "foolproofing" spirit that has proven so effective in Japanese manufacturing (Shingo, 1986).

RELEVANCE OF MANUFACTURING CONCEPTS TO DESIGN ENGINEERING

Design can be perceived as the manufacture of product concepts. Thus many of the rules that have been so successfully applied in the manufacturing arena can be adapted to improve the efficiency of the design process. The efficiency achieved in manufacturing is of course well known; a consequence of a strategy that has been elegantly stated by Toyota's Taiichi Ohno (1988):

> All we are doing is looking at the time line—from the moment that the customer gives us an order to the point when we collect the cash . . . (and then) reducing that time line by removing non-value-added costs.

Table 4.1. Parallels Between Manufacturing and Design Enhancements

Concept from manufacturing	Related issue in design
Continuous improvement	Self-examination and learning from past project experience
Poka-yoke	Foolproof design methodology at interdisciplinary boundaries Prototypes as total system checks
Elimination of waste	Minimizing waste caused by change Minimizing waste caused by unproven technology Minimizing non-productive activities
JIT	Resource support as a just-in-time operation
Worker flexibility and breadth	Designers with interdisciplinary experience
Flexible manufacturing	Flexible, automated design
Load balancing	Flexibility in breakdown of assignments—extra support at bottlenecks
Minimizing work-in-process inventory	Minimizing delay Maximizing utilization of past design Maximum utilization of supplier resources Focused design efforts

This brilliant, simple statement has significant relevance to engineering as well as manufacturing. Thus it is instructive to review some ideas derived from a critical rereading of Toyota Production System concepts. Some ideas from manufacturing that can contribute to improved engineering design productivity are compiled in Table 4.1.

The idea is certainly not to reduce engineering design to something akin to a relentless, rigid production line. The growth in engineering productivity must come not from restricting the engineer to a production line mentality, but rather from freeing the engineer from the non-value-added work that impedes effectiveness. Space does not permit much amplification of these parallels. As one limited example, consider the possibilities of flexible automated design as a counterpart to flexible manufacturing. Flexible manufacturing and group technology emphasize use of automated machine tools with minimum process planning and setup to produce a wide range of similar parts. Similarly, flexible automated design uses the power of computers substantially to automate the design and process planning of components. Not only does such a system emphasize efficient re-use of refined, standard methodology, but it also provides a strong driver to eliminate any redesign by making the designer fully aware of the available inventory of preexisting standard parts.

Vigorous efforts are being expended by Japanese firms toward this end (Sata, 1986; Ohara and Higashi, 1983; Yencho, 1991). Templin (1990) reports that Toyota seeks to create a large common database for the design of each basic component of an automobile, exploiting the generic similarity to prior designs. This cumulative database can contain historical experience emphasizing unique problems, warnings, and solutions, combined with basic design concepts that can be adapted and modified rapidly for the new application, followed by appropriate programs for the computerized analysis of important issues such as system performance predictions, formability evaluations, process plans, tolerances and robustness, special production considerations, and finite element analysis programs. Merely by making the modest changes in key dimensions that the new product requires, such a program can yield a complete component design and corresponding tool design and manufacturing process as well as incorporate the benefits of cumulative corporate experience and specialized knowledge.

As reported by Yencho, Toyota's system limits car body design to 12 basic curve shapes to simplify the CAD system needed to generate a corporate numerical control (CNC) program for manufacture of the body dies. He also reports that Mazda has achieved a 50% reduction in costs by developing a program of 1.5 million lines of code to reduce free-form shapes to steel dies in one step. Not the least of the many benefits of such programs is that, once debugged, they minimize loss due to routine errors.

CONCLUSIONS

This chapter has addressed questions concerning the productivity and time-responsiveness of the engineering design function and has examined some factors for improving the effectiveness and productivity of the engineering design function. Numerous opportunities for improvement exist—and each organization must adapt them to their peculiar circumstances. Progress will require a willingness to change in the spirit of total quality management, combining extensive self-examination and the implementation of many incremental improvements with the more extensive changes that fully exploit the powerful tools of the information age.

REFERENCES

Bauman, R. "CALS/CE: Mission Critical for Digital." *Aviation Week & Space Technology* 2, 1990, 1–12.

Bebb, B. "Quality Design Engineering—The Missing Link in U.S. Competitiveness." Pre-publication draft, Xerox Corporation, Webster, N.Y., March 1990.

Brown, J. S. "Research that Reinvents the Corporation." *Harvard Business Review* 69(1), January-February 1991, 102–111.

Clark, K. B., W. B. Chew, and T. Fujimoto. "Product Development in the World Auto Industry." *Brookings Papers on Economic Activity* 3, 1987, 729–781.

Fox, J. "Design Quality and Reliability Through Technology Readiness." Technical Report 52. Design Engineering Center, University of Missouri-Rolla, 1989.

Hammer, M. "Re-engineering Work: Don't Automate, Obliterate." *Harvard Business Review* 68(4), July-August 1990, 104–112.

Hauser, J. R., and D. Clausing. "The House of Quality." *Harvard Business Review* 66(3), May-June 1988, 63–75.

Ohara, M., and M. Higashi. "Integration of CAD/CAM Systems." *Automotive Body Engineering Computer & Graphics* 7(3–4), 1983, 307–314.

Ohno, T. *The Toyota Production System.* Cambridge, Mass.: Productivity Press, 1988.

Ross, P. *Taguchi Techniques for Quality Engineering.* New York: McGraw-Hill, 1988.

Sata, T. "Computer Integrated Manufacturing—Present State and Future." International Seminar on Factory Automation, Tokyo, December 5, 1986.

Shingo, S. *Zero Quality Control: Source Inspection and the Poka-yoke System.* Cambridge, Mass.: Productivity Press, 1986.

Sullivan, L. P. "QFD: The Beginning, The End, and the Problem In-Between." Quality Function Deployment, A Collection of Presentations and Case Studies. Dearborn, Mich.: American Supplier Institute, 1987.

Templin, R. "A Comparison of Methods in U.S. and Japanese Automobile Industry." Presentation to Manufacturing Symposium, Graduate School of Business, Stanford University, April 21, 1990.

Tichy, N., and R. Charan. "Speed, Simplicity, Confidence, An Interview with Jack Welch." *Harvard Business Review* 67(5), September-October 1989, 112–120.

Westney, D. E., and K. Sakakibara. "Designing the Designers." *Technology Review*, April 1986, 25–32, 68–69.

Wilson, E. "Product Definition: Assorted Tech-niques and Their Market Place Impact." *Proceedings of IEEE International Engineering Management Conference*, October 1990a, 64–69.

Yencho, S. "Observations of DFMA While in Japan." *Proceedings of the 5th International Conference on Design for Competitive Manufacturing*, Rochester, Mich.: Institute for Competitive Design, May 7–8, 1990.

5

THE DEVELOPMENT/MANUFACTURING INTERFACE: EMPIRICAL ANALYSIS OF THE 1990 EUROPEAN MANUFACTURING FUTURES SURVEY

ARNOUD DE MEYER

Competitive pressures force companies to improve the management of the interface between development and manufacturing so that they can shorten the time between product concept and market launch. Indeed, a delay in the introduction of a new product can significantly shorten the period during which the company can market the product, since the time of the withdrawal of the product is often determined by market factors beyond the influence of the company. As a consequence, a delayed market introduction can have a strong negative effect on the overall profitability of an innovation project.

Avoiding delays in market introduction calls for better control of the time between the first idea of the product and the launch of the product, or the time spent in developing product and process, and ramping up for production. In markets where product life cycles are shortening, the call for better control of the complete development cycle becomes even more important.

In this chapter, we will first demonstrate that controlling and shortening development cycle times is not only necessary, but can and should have positive effects on the quality and cost of the product development process. It is our hypothesis that controlling development time is in the first instance an issue of improving the interface between development and manufacturing. Technical tasks have to be solved, and taking away time from that problem-solving process is perhaps not efficient. But time can be gained in the transfer of technology from development to manufacturing, and in carrying out the problem-solving task in such a way that the job for the other party becomes easier.

After a short literature review on shortening design cycle times, we will analyze a database of 224 large European manufacturers in order to see what European manufacturers already apply with success.

SHORTEN DEVELOPMENT CYCLE TIMES AS A GOAL

The performance of product development activities in the firm has traditionally been measured against three types of standards: the quality of the design; the cost of developing and introducing the product to the market, i.e., the efficiency of the resource utilization during the product development process; and the lead time in which the design could be made ready for the market. These three types of performance measures are obviously interrelated.

Often it is argued that both costs and quality are inversely related to development and that costs in particular will increase more than proportionally if the development time is shortened. Graves (1989) concludes on the basis of a literature survey that 1% compression of the development time would result in a cost increase of 1.2% up to 2.0%.

But over the last few years it has been argued that this type of trade-off is unacceptable. Companies have to shorten their development time without giving in on product performance or development cost. The argument is not that the trade-off does not exist, but rather that most companies are far away from the efficient frontier. A vast improvement in development time can still be realized without negatively affecting cost or quality.

To support the need for shortening development lead times, three categories of reasons can be given. First there seems to be a general reduction of product life cycles for a vast range of products. This necessitates a more rapid development of products that will follow each other in the market. Obviously one could cope with shorter product life cycles by developing products in parallel. But such an approach can increase the complexity of the development process unless the products that are developed are chosen very carefully. Complexity is not the only consequence of parallel developments of successive products. Parallel development can lengthen product development times. Product managers of earlier products can indeed be tempted to try to include new features developed for a later product with the earlier product, causing in that way a further delay in the introduction of the earlier product.

The empirical support for shortening product life cycles is scant. But whether the product life cycles really have shortened, or have remained constant, is perhaps less important than the prevalent perception among managers that these product life cycles are becoming shorter or will shorten in the near future. Indeed, it is perception that influences behavior and creates the pressure to shorten development times.

The second category of reasons has to do with the increased variety of products offered in parallel to the customer. Increased product customization, i.e., increasing the number of options of base products through cost efficient design flexibility (De Meyer et al., 1989), leads to an increased need for improving the company's design and engineering capabilities. One of these capabilities is control over development lead time.

A third category of reasons can be derived from the set of ideas that can be broadly summarized as "time-based competition" (Stalk, 1988). The argument can be made as follows. Time is similar to inventories in manufacturing, the slack factor in development. Time is almost an equivalent of work-in-process in development. In manufacturing, the supporters of the just-in-time concept stress the need to reduce in-process inventories in order to expose problems and to improve the production process. Likewise reducing the slack in the development process will lead to an exposure of management problems in the way products and processes are developed. Reduction of lead-time in development should ultimately lead to better understanding and control over the development process and consequently to better product design quality.

A corollary of this third argument shows how shortening design cycle times can immediately improve the quality and the cost of the design of a product. If one can substantially reduce the development time of a product, one can start the development much closer to the date of market introduction. This reduces the need for forecasting of specifications and minimizes the risk of missing the real market needs. It is indeed easier to forecast market needs for a car two years before the date of introduction than six years before that date. Moreover, it reduces the need for development "rework," or the process by which some parts of the product have to be redesigned close to the date of introduction in order to take evolving market needs into account. Both quality of the final design and resource utilization during the development process can thus be improved by shortening the development time.

If one thinks carefully about these three clusters of factors, it quickly becomes clear that shortening development times is not a goal in itself. Control over the development time is in fact the real goal. Such control will ultimately lead to a better and cheaper product, which is what the customer is interested in. But it is probably true that most companies are not yet working at the efficient frontier where cost, quality, and development time are close to the optimum and have to be traded off with each other. Today, it still makes sense to focus on shortening development times, and to assume safely that this can be done without having a negative impact on cost and/or quality.

Several studies have been published on the shortening of development times: Clark, Chew and Fujimoto (1987), Clark (1989), Brockhoff and Urban (1988), Gold (1987), Gupta and Wilemon (1990), Mansfield (1988), De Meyer and Van Hooland (1990). All authors agree that the reduction of the development cycle is not a problem that can be solved by a development department in isolation. Product development requires a systems approach, and one needs to create at the very least a close collaboration among development, manufacturing, and marketing activities.

It is often argued that the greatest savings in the reduction of development times can be made in the interfaces between the different company functions involved in the development process. Indeed, it seems difficult to cut short the development tasks themselves. Specifications have to be defined by marketing, technical problems have to be solved by development, and processes have to be ramped up by manufacturing. But it seems that a great deal of time can be gained in the transfer of technology and information between functions, or in carrying out the different tasks in such a way that the results make the job easier for the other parties. Let us now first examine what is common knowledge about shortening the product development cycle.

HOW TO SHORTEN DEVELOPMENT TIMES: LITERATURE REVIEW

Earlier work on development time focused almost always on the reasons for delay. This reflects a rather negative attitude. Development time was considered to be planned and given. Management had to aim at avoiding delays. In a review paper on planning in R&D, Pearson (1983) makes no references to attempts consciously to shorten development times. Though he points out that some authors describe the lack of resources as the most important reason for delays, he also reports on studies from the early seventies in which it was shown that one third of time lost in development could be attributed to lack of manpower resources, one third to problems with deliveries of supplies and services and one third to unexpected technical difficulties. Let us first take the issue of lack of resources.

Gupta and Wilemon (1990) report on the basis of 38 interviews that 42% of their interviewees cited lack of resources as an important reason for the delay of a project. These studies fit the context of the time-cost trade-off to which we referred in the previous section. Indeed, if lack of resources leads to delays, one could infer from this that delays can be eliminated or that development times could be shortened if the resources available to the project would be increased. But in an earlier study (De Meyer and Van Hooland, 1990), we found no indication whatsoever that a higher availability of resources would contribute to shorter development times. Indeed, neither the resources spent on R&D (as a % of sales) nor the ratio of process engineers to total number of employees could explain the success of a company in improving its speed in introducing new products in the markets. Brockhoff and Urban (1988) have reported on a study of 31 projects in an electrotechnical company. In their analysis of the delays they distinguish four main categories: technical difficulties during the development phase (explaining 53% of the delay), failure or late delivery of components (30% of the delays), problems during the prototype construction (10% of the delays), and changes in concept definition during the development stage (7% of the delays). Again there is no real reference to a lack of resources. A careful reading of Gupta and Wilemon (1990) shows that the lack of resources on which they report is really a symptom for underlying shortcomings in the new product development process, such as insufficient support from top management, or too many inexperienced people working on the project. Only a few of their interviewees noted that resources were taken away from important projects that substantially slowed their progress. The first conclusion seems thus to be that simply throwing resources at the development/manufacturing interface is not the solution to improve development performance. Managerial actions seem to be required.

Both Wilkes and Norris (1972) and Brockhoff and Urban (1988) came to the conclusion that about 30% of the delays are due to vendors. The idea that vendors and suppliers play an important role in development time seems to be very important.

But not only quality and on-time delivery of parts and services are important. In many cases suppliers will go a step further and carry out part of the development and design task. They not only produce goods and services. They also produce ideas for product development. Imai, Nonaka, and Takeuchi (1985) and Clark (1989) have documented the role of vendors in the design of assembled products with respect to five in-depth case studies in Japan, and for the world's automobile industry. Both studies stress the need for an interorganizational network in which vendors and suppliers carry out part of the development task for the main contractor. In De Meyer and Van Hooland (1990), improved vendor relations were mentioned as the single most important factor explaining the increase in speed of introducing new products. Clark (1989) mentions that involving suppliers does not automatically lead to a reduction in development times. Indeed, there is no guarantee that a supplier is actually more efficient or faster than the main contractor. In many cases one could assume that a small supplier does not have both organizational and technical capabilities to carry out a major development task. Suppliers can become an asset to the development process, on condition that they form part of a carefully nurtured network in which information flows freely and where the main contractor helps in the upgrading of the subcontractor's development capabilities.

Brockhoff and Urban (1988) structure the actions one can take to influence the length of the development project around the concept of a better management of the subprojects. Apart from a good definition of the subprojects and a careful control of the subprojects on the critical path, they mention the need to carry out parallel projects; an evolving definition of the subproject's definition; and they advise defining projects as independently as possible. They insist on the organizational measures one has to implement. They emphasize the qualifications of the project leader and personnel, the active management of the degree of acceptance of the deadlines, and the measures taken to meet those deadlines.

It is also suggested that early involvement of other functional groups such as manufacturing may lead to a reduction of the technical problems encountered during the development phase. This early involvement can take the form of job rotation, regular joint review meetings, seminars, joint customer visits, social interactions, and physical proximity of the workplaces (Gupta and Wilemon, 1990; Moenaert and Souder, 1990).

Using the same organizational approach, Fruin (1988) describes a clinical study of Toshiba's "development factory." He argues that Toshiba has two types of factories: one for normal mass production and one for developing and ramping up of new products. In the analysis of the characteristics of this development factory, Fruin stresses the importance of aspects of human resources (feeling of community, liveliness of the factory, high specialization by function and product area) and on organizational characteristics (centrality of the factory in the organization, high leverage of the company's skills in the factory, large size of the factory). New forms of human resource management and creative ways of looking at the organization of the R&D/manufacturing interface seem to be indicated.

Bringing a product from conception to market introduction is a very data-intensive process. It has become common practice to describe product innovation as a process of data-gathering and creative data-processing. In this context two elements emerge that can help to shorten the development time: parallel problem-solving and data integration.

Clark and Fujimoto (1989) assert that the shorter development times they observed in some automobile companies can be explained to some extent by parallel or simultaneous engineering. The distinction between parallel and sequential problem solving in development is not new. It was described by Abernathy (1971) and has been stressed as a critical element of success of the Japanese new product development process by Imai, Nonaka, and Takeuchi (1985). The idea of parallel or simultaneous development is quite simple. In order to shorten development times one should replace the sequential execution through which the development goes by a partially parallel execution of the development stages. By doing so, the duration of each subproject or stage in the project is not necessarily reduced, but the overall length of the project is reduced. The principle of overlap between product and process engineering is simple (Ettlie and Reifeis, 1987), but its practical application creates many organizational problems. Indeed, parallel engineering requires collaboration, interaction, and coordination between teams with different functions. Interfunctional teamwork and early interfunctional conflict resolution become conditions necessary to succeed with parallel engineering.

The second element that is related to innovation as an information-intensive activity is the integration of the data used and generated during the innovation process. The data used in innovation and in particular shared by manufacturing and development can take many forms: geometric representations, algorithms, product and supplier specifications, scientific literature, customer data, process data, etc. In order to improve development time it would seem appropriate that all parties involved in the development process have equal access to a convenient database about materials, processes, equipment, tools, market data, and other relevant information. In earlier days, all of this information was available on some hard copy system, but more recently most of this data, if not all, is available on computers. Today's technology enables us to integrate, at least theoretically, all these data in one system accessible by all functions. Some vendors today offer computer-integrated management systems that make all databases totally transparent. Investment in this type of hardware should at least create the possibility to shorten development times.

Beyond the integration of information systems, one can imagine the automation of the design process contributing to shorter development cycles. Design process automation, such as CAD, CAM, CAD/CAM, desktop graphics, supercomputers, and other tools that automate aspects of the development process are often proposed by the vendors to be a panacea to shorten development times. Theoretically they have that possibility, but empirical evidence shows that in practice few CAD/CAM installations have actually led to shorter development times (Adler, 1989).

From the perspective of manufacturing, it seems that action programs aimed at shortening development times will be more effective if they are absorbed by manufacturing organizations that have made investments in other types of programs aimed at getting the manufacturing function under better control (Hayes, Wheelwright, and Clark, 1988). These programs are often summarized in statements such as "world-class manufacturing" and include JIT, statistical quality control, simplification of the production process, and enhancement of manufacturing flexibility. This seems to have validity: companies that pay attention to the technological function up to the point where they try to transform manufacturing into a competitive weapon will probably be able to get rid of the slack in the development system.

To summarize, one can say that though the literature recognizes the importance of resource availability, it stresses the need to accompany or replace increases in resources with managerial action. This managerial action can focus on better relations with suppliers; better project management; both from a technical and an organizational viewpoint; integrated forms of organizations; a parallel form of problem solving; deployment of computers in order to increase data integration; automation of the design process; and a manufacturing environment conducive to innovation.

Applying all these managerial actions at once not only seems to be a Herculean task, but could as well disrupt the whole organization. In an earlier paper (De Meyer and Van Hooland, 1990), we argued that the portfolio of actions described in previous paragraphs cannot and should not be implemented simultaneously. Instead, they should perhaps be viewed as first- and second-order activities. On the basis of that earlier study, it seems that a close development/manufacturing collaboration requires an up-front commitment to shorten development times, an investment in different (non-Taylorian) management-labor relations, and the application of a systems view to the innovation process (i.e., involve customers and vendors in the development tasks). The environment created in that way will become a fertile ground for introduction of simultaneous engineering, interfunctional project teams, integration of information systems, and design automation.

SOME EMPIRICAL OBSERVATIONS

In order to explore some of the assertions in the literature review, we will use the data gathered in the European Manufacturing Futures Survey.[1] It is not the purpose of this empirical analysis to test the conclusions of the previous section. Rather, we would like to understand to what extent a selected group of manufacturers already applies some of the techniques proposed in the literature.

The survey is a biennial survey of large European manufacturers. The target respondents are senior manufacturing managers. The constraints within which we have to interpret the results are obvious: the survey is limited to manufacturing managers of large, established companies. The managers who answer the questionnaire have a high level of seniority in the company. Consequently they may have a view that goes beyond the boundaries of the manufacturing function.

How can the data be used? In the survey, we asked the respondents to construct an index of how their business unit improved from 1987 to 1989 on a number of performance indicators (assuming that the performance in 1987 was equal to 100). One of these performance indicators is the speed of new product introduction. These indicators seem to be valid when compared to other independent indicators measuring the same type of performance.

In a second section we presented to the respondents a list of 26 action plans. For each of them, we asked them to check off whether these action plans had been given a significant emphasis over the last two years, and to rate on a seven-point scale what emphasis would be placed on these action plans over the next two years. We also asked the respondents to rate on a seven-point scale the payoff of these programs. In a separate section we asked for a number of characteristics about the development process, such as the degree of overlap between product and process engineering, the utilization of CAD, the integration of CAD and CAM, and the involvement of suppliers in the development process. Both action plans and characteristics of the development process together cover most of the items discussed in the literature review.

The logic we will use is the following. First, a description of the absolute results will be given. Second, the performance indicator will be used as a dependent variable, describing the effectiveness of the development/manufacturing interface. The characteristics of the innovation process and past and future action plans will be used as explanatory variables of the performance improvement. Obviously one has to be careful. All answers are perceptions and self-ratings. Moreover, the distribution of the answers for some of the questions is highly skewed. Assuming normality of the

Table 5.1. Sample Characteristics
(Medians except for sales)

Average annual sales (in ECU)	32207.50 ECU[a]
Pre-tax return on assets (%)	13.4
Net pre-tax profit (% of sales)	7.5
R&D expenses (% of sales)	3.0
Unit growth in sales (%)	13.0
Market share primary product (%)	25.5
Capacity utilization (%)	90.0
Number of plants	2
Number of employees	730
Number of direct mfg employees	350
Number of indirect mfg employees	140

[a] ECU (European currency unit) is about U.S. $1.30.

results does not seem appropriate. Both t-tests and Mann-Whitney tests were used for this reason.

The Sample

In the first quarter of 1990 we received 224 answers from business units located in 14 European countries. One hundred eighty-five answers came from companies in European Community countries. The characteristics of the sample are described in Table 5.1.

The variety of industries represented in the sample is quite large and covers many three-digit SIC codes. The sample is somewhat biased in favor of basic industries, which include all base, fine, and specialty chemicals; petrochemicals; glass; and steel and nonferrous metals and their transformation. This bias may actually reflect where some of the strengths of European industry can be found.

Description of Some Relevant Results

The average European business unit in our sample does not feel too good about its performance with respect to the introduction of new products. Though the respondents report on the average an improvement of 8.1% over the period 1987 to 1990, this is the third-lowest improvement rate out of a list of 14 performance indicators, and considerably lower than the reported 28% in profitability or 15% in overall quality. One hundred three respondents reported a deteriorated or constant performance in speed of new product development/design change. This weakness is reflected in the action programs. Indeed, those action programs directly related to the development of new products or to the development/manufacturing interface have not been emphasized heavily over the last two years. Out of a list of 26 action programs that the respondents could check off, programs such as CAD, design for manufacture, and value analysis are only moderately to lowly emphasized (Table 5.2). Let us take the example of CAD. Only 37% of the respondents indicated that they had significantly emphasized investments in CAD over the last two years. Within the list of 26 action programs, CAD was only the fifteenth most emphasized program. Design for

Table 5.2. Ranking and Emphasis on Action Programs Related to the Development/ Manufacturing Interface

Action program	Percentage who checked	Rank[a] importance	Rank[a] Pay-off
CAD	37	15	13
Value analysis and product standardization	23	22	15
Interfunctional teams	36	16	9
Quality function deployment	52	5	6
Development of new processes for			
Old products	35	17	3
New products	40	11	5
Design for manufacture	15	25	21

[a] Out of 26 items.

manufacture was the second least emphasized program in the total group of 26 action programs.

Moreover, if we examine the evaluation of the payoff of these programs, it seems that the lowly emphasized programs did not get an enthusiastic evaluation by those who applied them. CAD was valued to have the thirteenth highest payoff. Design for manufacture was only 21st in the ranking of payoffs of action programs. Of course, one has to take into account that some of these programs have just been started and can take a long time to show a reasonable return.

Quite a few of the respondents already use both CAD and CAM (Table 5.3). The degree of integration is, however, not so high. Almost two-thirds of the respondents who have both CAD and CAM have a one-way integration, i.e., do a downloading from design to manufacturing. But only one-quarter of these respondents (i.e., about 12% of the total sample) indicate that there is some form of two-way communication.

The overlap between process and product engineering is less than might be expected (Table 5.4). Process design starts when 50% of the product design has already been done (3.8 on a 7-point scale).

The average number of people in manufacturing who have design experience and vice versa is quite high: about one-quarter of them have worked for more than three months in the other function. But the results are very skewed. The median is

Table 5.3. Degree of CAD/CAM Integration

	%
Respondents having both CAD & CAM	48
Respondents who have implemented	
Automatic downloading from design to mfg	62.5
Design rules for producibility	53.4
Two-way communication allowing mfg to send data to design	23.9
Automated process planning	39.8

Table 5.4. Organizational Measures

Overlap product-process engineering	3.8[a]
Manufacturing personnel with design experience	26.9%
Design personnel with manufacturing experience	26.5%
Product determines process	5.0[b]
Process determines product	4.7[b]
Planned joint activities with vendors	81%

[a] On a scale from 1 to 7—1 = no overlap, 7 = simultaneous development.
[b] On a scale from 1 to 7—1 = no influence, 7 = major influence.

12% and 10%, respectively. It appears that product tends to determine the process specifications a little bit more than vice versa. Finally, more than 80% of the respondents intend to develop nonexclusive joint activities with vendors over the next two years.

With respect to suppliers Table 5.5), it seems that about 50% of the products purchased are proprietary products from the supplier. About 30% of the products are designed by the company and subcontracted. About 15% of the components are designed by the subcontractors. It has to be mentioned that only about half of the respondents answered this question, and the results should be treated with caution. There is no significant industry influence on the answers to this question.

Influence of Manufacturing on the Development/Manufacturing Interface

What kind of action programs have been applied in order to improve the development/manufacturing interface? We will take the estimation of the performance improvement in speed of new product development as the dependent variable and relate these to the emphasis with which certain action programs have been implemented (see Table 5.6).

For the estimation of the improvement of the speed of new product development, the high performers were defined to be the ones that score higher than average (108) plus half a standard deviation ($s = 17$). The low performers were defined as those companies that score lower than the average. A similar procedure to separate high and low performers was used in other studies (Allen, 1977).

We also could have taken the companies whose performance in product development speed remained constant. We preferred to include only those who report lower scores. In that way, we correct for some of the reported "wishful thinking"

Table 5.5. Collaboration with Suppliers (n = 119)

Relative proportions of components and subassemblies by type

Type	As a % of units	As a % of value
Proprietary products of suppliers	51.2	47.2
Designed by company and produced by supplier	28.8	30.2
Designed and produced by supplier	13.1	15.2
Others	6.8	7.2

Table 5.6. Activities Stimulating the Development/Manufacturing Interface (Complete Sample n = 224)

	p-value (t-statistic)	Low performers	High performers
Future investments in action programs			
Worker training[a]	.02	5.2	5.7
Management training	.07	5.1	5.5
Supervisor training	.09	5.3	5.6
CAD	.05	4.3	5.0
Interfunctional teams	.07	5.0	5.4
Value analysis	.09	4.1	4.7
New processes for new products	.04	4.8	5.4
Just-in-time	.07	4.7	5.2
Design for manufacture	.03	4.0	4.9
Close plants	.03	4.0	4.9
Quality circles	.004	3.9	5.1
Improve production and inventory control systems	.02	4.6	5.4
Hiring outside skills	.08	3.8	4.5
Past investments in action programs			
Integrate information systems across functions (%)	.01	51	28
Just-in-time (%)	.04	35	52
Characteristics of the development process			
Suppliers proprietary products in total supplies (%)	.07	56	41
Products designed by main contractor in total supplies (%)	.01	30	49
Overlap product/process engineering[b]	.07	4.0	3.4

[a] All thirteen items in this section were rated on 7-point scales; 1 = little emphasis; 7 = great emphasis.
[b] One (1) means 100% overlap, while seven (7) means a completely sequential process (i.e., process design starts after product design is completed).

improvement. Having separated the sample in these groups, t-tests and Mann-Whitney tests were used to measure whether the groups differed in performance improvement. Simple tests were preferred over a discriminant analysis since the distribution of the answers to some of the independent variables did not allow us to apply multivariate procedures.

The high performers intend to invest in the near future in worker, management and supervisor training, CAD, interfunctional teams, value analysis, new processes for new products, just-in-time, design for manufacture, quality circles, and improved production and inventory control systems. They also plan to close plants and hire outside skills. When it comes to past investments, the high performers spent less for integration of information systems across functions, but spent more on just-in-time programs. With respect to characteristics of the product development process, high performers used proprietary products of suppliers but designed relatively more of the components themselves. They also did more parallel process and product development.

It is of course equally interesting to mention some of the programs that did not make any difference at all and that are consequently not mentioned in the table. These included intention to have joint activities with vendors, a higher proportional investment in R & D, the proportion of indirect manufacturing personnel to direct manufacturing personnel (as an indicator of process engineering capabilities), the number of drawings generated by CAD, or the degree of CAD/CAM integration.

DISCUSSION AND CONCLUSIONS

The group of programs that make a difference is quite small. Several explanations can be found for this. There are, of course, the inherent limits of the data collection method. Large sample questionnaires can often only measure very general variables and use proxy measures. Moreover, some programs may take more than the two years covered by the questionnaire to have an impact. It has been argued that some programs may even have a negative impact in the short run while having a very positive impact on productivity improvement in the long run (De Meyer and Ferdows, 1990a). In this analysis this could be the case for the interfunctional electronic data integration in the electronics industry.

Having set these constraints, the main result seems to be that the European companies that have been successful in shortening their development time have done so mainly by implementing just-in-time principles, by increasing the overlap between process and product design, and by designing a higher percentage of their own products in-house. Their success is inversely related to integrating information systems across functions. Successful companies intend to invest in training at all levels, in continuous process improvement, and in infrastructural programs such as design for manufacture, production and inventory control value analysis, and interfunctional teamwork.

When we compare this to the other studies mentioned in the literature review, we see some support for the hypothesis that investments in computer integration of information have not yet paid off for European manufacturers. They appear to be successful by doing more parallel problem-solving and using suppliers in the development process, but more as suppliers of proprietary products than as partners in the development process. Organizational measures are not really mentioned. This is in line with a more general analysis of European industry (De Meyer and Ferdows, 1990b) where it is argued that over the last few years European manufacturers have tried to improve their global competitiveness mainly through structural changes in their production function. Having partially achieved these structured changes, they are now prepared to open up the manufacturing boundaries to integrate manufacturing with other functions in the company.

It seems that European industry is only at the start of the application of some of the principles outlined in the literature. A lot of emphasis has probably been put on the structural side of the investments. Too little attention has been paid to the organizational or infrastructural aspects of the innovation process. The results are not very positive either: on the average, European industry does not shine in its development speed, and managers know it. It seems, however, that a healthy reformulation of the manufacturing efforts that can lead to shorter development times is neverthe-less under study.

NOTE

1. The European Manufacturing Futures Survey is part of a larger research project, the Global Manufac-turing Futures Survey. This project was started in 1981 at Boston University and has been carried out since 1983 at INSEAD (Europe); Waseda University (Japan); and Boston University (United States).

REFERENCES

Abernathy, W. "Some Issues Concerning the Ef-fectiveness of Parallel Strategies in R&D proj-ects." *IEEE Transactions on Engineering Man-agement* EM-18(3), 1971.

Adler, P. "Integration of CAD/CAM." *Special is-sue of IEEE Transactions on Engineering Man-agement*, 1989.

Allen, T. J. *Managing the Flow of Technology.* Cambridge, Mass.: MIT Press, 1977.

Brockhoff, K. and C. Urban. "Die Beeinflussung der Entwicklungsdauer." *Zeitschrift fr Betriebswirt-schaftliche Forschung.* Special is-sue 23, 1988, 1–42.

Clark, K. B. "Project Scope and Project Perfor-mance: The Effects of Parts Strategy and Sup-plier Involvement on Product Development." *Management Science* 35(10), 1989, 1247–1263.

——— and T. Fujimoto. "Overlapping Problem Solving in Product Development. In *Managing International Manufacturing,* ed. K. Ferdows. Amsterdam: North Holland, 1989, 127–152.

———, W. B. Chew, and T. Fujimoto. "Product Development in the World Auto Industry." *Brookings Papers on Economic Activity* 3, 1987, 729–771.

De Meyer, A. and K. Ferdows. "Influence of Manufacturing Improvement Programs on Per-formance." *International Journal of Operations and Production Management* 10(2), 1990a, 120–131.

——— and ———. *Removing the Barriers in Manufacturing.* INSEAD Working Paper No. 90/73/TM, 1990b.

——— and B. Van Hooland B. "The Contribution of Manufacturing to Shortening Design Cycle Times." *R & D Management* 20(3), 1990, 229–239.

———, J. Nakane, J. G. Miller, and K. Ferdows. "Flexibility: The Next Competitive Battle." *Strategic Management Journal* 10(2), 1989, 135–144.

Ettlie, J. and S. A. Reifeis. "Integrating Design and Manufacturing to Deploy Advanced Manu-facturing Technology." *TIMS-Interfaces* 17(6), 1987, 63–74.

Fruin, M. "History, Strategy and the Development Factory in Japan." EAC Working Paper, IN-SEAD, 1988, Fontainebleau, France.

Gold, B. "Approaches to Accelerating Product and Process Development." *Journal of Production and Innovation Management* 4(2), 1987, 81–88.

Graves, S. B. "The Time-Cost Trade-Off in Re-search and Development: A Review." *Engineer-ing Cost and Production Economics* 16(1), 1989.

Gupta, A. K., and D. L. Wilemon. "Accelerating the Development of Technology-based Products." *California Management Review,* Winter 1990, 24–44.

Hayes, R., S. G. Wheelwright, and K. B. Clark. *Dynamic Manufacturing.* New York: The Free Press, 1988.

Imai, K., I. Nonaka, and H. Takeuchi. "Managing the New Product Development Process: How Japanese Factories Learn and Unlearn." In *The Uneasy Alliance: Managing the Productivity-Technology Dilemma,* eds. K. B. Clark, R. H. Hayes, and C. Lorenz, Boston: Harvard Busi-ness School Press, 1985, 337.

Mansfield, E. "The Speed and Cost of Industrial Innovation in Japan and the United States: Exter-nal vs. Internal Technology." *Management Sci-ence* 34(10), 1988, 1157–1168.

Moenaert, R. and W. Souder. "An Information Transfer Model for Integrating Marketing and R&D Personnel in New Product Development Projects." *Journal of Product Innovation Man-agement* 7(2), June 1990, 91–107.

Pearson, A. "Planning and Monitoring in Research and Development." *R & D Management* 13(22), 1983, 107–116.

Stalk, G. "Time, the Next Source of Competitive Advantage." *Harvard Business Review* 66(4), 1988, 41–51.

Wilkes, A., and K. P. Norris. "Estimate Accuracy and Causes of Delay in an Engineering Research Laboratory." *R & D Management* 3(1), 1972, 35–46.

6
MODULAR DESIGN AND THE ECONOMICS OF DESIGN FOR MANUFACTURING

GORDON V. SHIRLEY

During the past two decades there have been a number of fundamental trends that have affected manufacturing-based industries. Among the more significant of these have been a rapid rate of development of new materials and material processing technologies, a large increase in the number of international players, and a growing integration of global markets. Together these changes have had a significant impact on the nature of competition. As a result, manufacturers in the affected industries have had to develop new approaches to remain competitive.

As the number of global competitors has increased, competition based on cost and lead times has intensified. In some instances, new competitors have found it possible to enter existing markets by exploiting differentials in factor costs to offer comparative products at lower prices. More often than not, however, entry has come only when the products offer distinctive capabilities, possibly targeted at specific segments in the major markets (Clark, 1985). To stave off this type of entry, it has been necessary for incumbents to eliminate potential entry points by developing a range of products targeted at different segments of the market (Lancaster, 1979; Schmalensee, 1978; Eaton and Lipsey, 1976). To execute this strategy, it is necessary to develop production systems capable of producing a wide range of products cost-effectively. The opportunities for competing on the basis of product differentiation have been enhanced by the rapid pace of developments in the underlying material sciences. Often these developments make it possible to offer substitute products that provide a more attractive mix of features at lower costs, thus making existing products obsolete. To cope, manufacturers have had to focus on the development of methods to reduce product design and development time in an effort to bring new generations of product to market more rapidly (Clark and Fujimoto, 1989; Shirley, 1987).

Much of the current interest in design for manufacturing, design for assembly, and methods for managing the design-manufacturing interface results from an awareness of these developments. The current interest in these methods is shared by practitioners as well as by researchers in the fields of engineering, manufacturing, and production management. Different groups have focused on different sets of requirements. Thus, some researchers have focused on the changes that are required in the design of products to reduce production costs, setup requirements, and lead times (Boothroyd and Dewhurst, 1987; Lee and Hight, 1989). Another body of work has focused on design of "variant" products and on improving manufacturing flexibility as a basis for producing a larger mix of products (Bracken, 1984; Bracken and Isolia, 1984). Other researchers have focused on methods of improving creativity in the product development process.

The view taken in this chapter is that a suitable response to the changes that have occurred requires that attention be paid to all of the different requirements simultaneously. It requires careful attention to the processes by which products and manufacturing processes are designed. A useful point of departure in the effort to understand the changes that are necessary in design procedures to cope with the new competitive requirements is to examine the nature of design processes and the coordination problems that arise in design. This is the topic of the first section. In analyzing the design process, special attention is focused on the methods by which formally and informally defined constraints are managed in the design process. The design rules that form a central part of the design for assembly/design for manufacturing methodology are analyzed in terms of the constraints they impose on designers during a product (or process) development exercise (Boothroyd, 1987; Redford, 1984; Poli, Graves, and Groppetti, 1986).

In discussing the design process, special attention is also placed on the procedures used to decompose large, complex design problems into smaller, more manageable sub-problems. The approach to decomposition is shown to have a significant impact on the resources consumed and the duration of design processes. In the second section, the concept of "modularity" in design is discussed; and its impact on the overall design time, on the ability to make changes to the design of a product or process as it is being developed, and on the ability to use knowledge generated in one aspect of the design in solving other related sub-problems is examined.

Examples from practice discussed in the third section provide concrete illustrations of the concepts described in the second and third sections of the chapter. These examples are drawn for the microcomputer industry. While we discuss the concept of modularity as it has been applied in the design of both the product and the production process, comparatively more attention is focused on its application in the latter situation, where the application of the concept is comparatively more novel. The use of the modular approach to design in this context underlines the generality of the concept.

There is another rationale for analyzing the design of the production system in detail. By doing so we obtain some interesting insights into the economics of designing products and processes for ease of manufacturing. These insights are discussed in the fourth section. The outline of a simple model is briefly described to illustrate how these insights can be incorporated in decision aids designed to enhance managerial intuition of the design for manufacturing process.

THE PROCESS OF DESIGN

Design is the process of specifying the characteristics of an artifact in sufficient detail to allow it to be fabricated and to have confidence that its performance is acceptable relative to a set of a priori goals. The fundamental task faced by the firm in the design process is to link in concept the needs of the market with the possibilities of available technologies. Typically this is an incremental and iterative process involving two parallel complementary activities that converge in the final solution (Shirley and Eastman, 1990).

The first of these activities begins with the general performance requirements of the intended user(s) of the product and deduces from them the desired system performances that are thought to achieve them. This process of deduction translates broad functional objectives (performs computations quickly, consumes less power, low maintenance, etc.) into properties of the object that can be realized within the design and engineering process. The logic of these ascriptive translations is generally weak and is often determined by what designers believe is achievable.

The other activity works from the known properties of materials and composes the object by a process of inductive reasoning and analysis. In this way a description of the object is developed from the known properties. Tolerances may define the bounds on property values allowed for geometric or other types of properties.

It is the desired functionality of the product that determines which properties of the product are relevant. In most cases, a product is required to perform multiple functions, and for analyzing each of these it may be necessary to develop different representations of the product. By a representation we mean an abstract model of the system in which certain details are omitted. The multiple dimensions of performance of the product, characterized in multiple representations, is usually the responsibility of a diverse set of design experts. These experts understand the relationship between physical properties of the object and performance and are able to evaluate candidate designs in their performance domain.

With multiple functional requirements, most products are highly complex. A usual method of dealing with complexity is to decompose the design in a (quasi)-hierarchic manner into sub-problems, which are in turn subdivided into smaller problems, and so on. The solution to each sub-problem at the lowest level consists of a physical object or component, and most engineering designs are composed of many component parts. Assessment of performance requires accumulating the performance of individual parts into the performance of higher-level aggregations of components. In the end, the descriptions generated for a design provide a logical structure between individual material properties of the elemental components and the desired high-level systemic performances that are deemed acceptable based on the analysis of market needs and desires.

The overall structure of design descriptions is an AND-OR graph (Marple, 1961; Eastman, 1988). The AND nodes define the technologies selected for use and their necessary components. The OR nodes are the local alternative realizations of the various technologies. The design process, viewed abstractly, involves defining this graph; that is, in defining the AND nodes, and searching among the relevant OR branches for a combination that best meets the ascriptively defined goals.

Design Constraints

The task of specifying the goals for the object under design can be likened to the task of establishing a set of constraints within which operational performance should lie given the expected range of environmental conditions. These kinds of constraints are referred to as performance constraints (Eastman, 1988).

In analyzing the alternate realizations of a technology to find the combination that best meets the ascriptively defined goals, certain compositions will be meaningless because physical reality does not allow them to exist. Physical objects impose strong constraints on the combination of functional properties they may share. Physical definitions impose equivalence across sets of variables, such as mass-density-volume or volt-amp-watt. In other instances the geometry of the object imposes other restrictions. These definitions and relations impose restrictions on the combination of performance values that are possible in a single design object. Even where some combinations of performance capabilities are theoretically possible in an object, no known means of fabricating such an object exist. They impose "realizability" constraints in the design process.

Another set of constraints deals with the relations between pairs of entities. While certain entities themselves may be meaningful, their combinations are not. Flow properties between adjacent pairs of objects must be consistent, and unrealizable conditions exist if pairs of solid objects occupy the same location in space at the same time. Constraints between objects are called "compositional constraints."

Constraint management in practice. Constraints are a general characterization of design knowledge. In most domains, formal definition of applicable constraints has not been attempted. Some efforts have been made to represent design problems as a sequence of constraints and to develop optimization methods for analyzing these problems. Typically these have been applied to single-level design problems, rather than multilevel complex products such as the ones with which we are concerned. In design processes the iteration that occurs between formulating in concept the needs of the market in terms of a set of realizable functional requirements and the specification of the product for meeting these requirements is often a reflection of the method used to deal with these constraints. A common method is to begin with many of the constraints relaxed. The goal is to find and develop a (set) of feasible solutions that meet the most basic functional requirements. Progressively, constraints relating to cost and additional functionality are imposed. Typically, these require changes in some aspects of the design that may lead to changes in many other decisions that have been made previously. Because designers with expertise in one domain may be unfamiliar with the impact of a change in the physical object on performance in another domain, the design process involves a series of iterations between different functional experts, converging, often slowly, to a final solution. The pattern of changes is often ill-defined in practice, often determined by organizational culture.

The length of time involved in converging to a design solution is therefore often a reflection of the informal methods of dealing with constraints in the design process. Often only very weak realizability constraints relating to manufacturing are imposed. Feasibility in manufacturing is often the initial concern. The lack of subsequent attention to "tightening" these constraints to eliminate costly fabrication procedures is in part a reflection of (a) the lack of expertise concerning current manufacturing processes on the part of design departments and (b) the length of time required in converging to a design solution, which may curtail additional efforts further to improve the solution by imposing tighter constraints concerning manufacturing.

DFM/DFA and design constraints. With this view of the design process, design for manufacturability, particularly as discussed in the engineering literature, can be viewed as an attempt to define in a formal manner sets of realizability constraints for specific classes of processes. As noted, realizability constraints considered by designers have typically been concerned primarily with the physics of the problem. Where constraints imposed by the manufacturing capabilities of the firm were considered, concern has usually focused on eliminating infeasible solutions. DFM imposes stronger constraints as it relates to manufacturing by establishing not only what is feasible, but what is cost-efficient to achieve, and by exclusion, what is difficult to fabricate.

In a similar vein, design for assembly can be construed as an attempt to define a set of compositional constraints that limits the feasible set of compositions of objects to be considered. Thus designs that call for an object to be inserted from the bottom up into another object may be excluded because this represents a costly assembly operation.

The DFM/DFA exercise as presented in these studies has two goals. The first is to eliminate costly design solutions and thus reduce the cost and lead time of products. The second objective is to reduce the amount of iteration that occurs between the design and manufacturing stages of the development process. Because there is no general method of dealing with constraints in the design process, the imposition of additional constraints can often result in a greater amount of "search" time in the engineering phase of the process.

In general, there is a need to understand how the process of design can be managed in order to achieve the other goals identified above. In this context, the concept of modularity in design is presented as a method that is complementary to conventional DFM/DFA efforts and provides some insight on how the combined set of manufacturing requirements can be jointly achieved.

MODULARITY

While most complex design problems are decomposed hierarchically, not all hierarchical decompositions lead to reductions in the complexity of the design problem. In general, decompositions in which the sub-problems that can be understood individually are preferable to ones in which a change in the solution to one sub-problem affects aspects in other sub-problems. In the latter case, the design process is difficult and time-consuming because a small change to improve one part of the design has to be accompanied by many simultaneous compensatory changes elsewhere. Decompositions in which a large problem is split into a collection of parts that are as nearly independent of one another as the task allows are referred to as "modular" decompositions, and the separate sub-problems defined in this way are referred to as "modules" (Parnas, 1971, 1976; Stevens, Myers, and Constantine, 1974). Modularity does not, of course, forbid weak interactions between the different modules in a task, but it implies that, to a first approximation, the overall organization is modular.

Modularity is a fundamentally important concept in the design of complex artifacts and has an important bearing on the set of manufacturing requirements described at the outset in several important ways. The specific features of a modular design and the manner in which they impact these requirements can be summarized briefly as follows.

Independence of Design Sub-Problems

When a design problem is decomposed such that the level of interdependence between sub-problems is low, it allows designers to proceed with different aspects of the design in parallel. Parallelism in design is important because it leads to reduced overall design times. Each sub-problem may itself be complex, involve effort by different functional specialists, and itself be subject to further decomposition. If each design engineer has to know the details of the decisions being made by each of the other designers in the project, the total amount of information that has to be communicated grows proportionately with the square of the size of the problem measured in terms of the number of designers involved, or the number of sub-problems to be solved. If however, the design problem is sub-divided progressively in such a way that the designers working on each sub-problem need detailed information only about the decisions being made by designers working on the same sub-problem and aggregate information about the other sub-problems, the total amount of information that must be transmitted will grow only slightly more than proportionately with the size of the problem, and the amount of communication required by each engineer will remain nearly constant (Simon, 1960). Modularity therefore simplifies the information-processing requirements in a design project. With the complexity of each sub-problem controlled in this way, it is generally possible to impose stronger realizability and compositional constraints without making the problem unduly difficult.

When products are modular in structure, it generally allows for more efficient customization. This is particularly true when each module of the design is concerned with the performance of a different functional requirement. Classes of products, each offering different levels of functionality and/or performance levels in each functional area can be developed by progressively modifying different modules of the product, without the need to make substantial changes in others. Modularity of this form therefore allows the "spatial requirements" to be efficiently addressed.

Abstraction, Aggregation, and Generalization of Information

At each level of a hierarchically decomposed problem, design involves identifying candidate technological solutions and evaluating their performance relative to the goals that have been established a priori. The principle of "abstraction" is of fundamental importance in performing these evaluations. By abstraction we mean the ability to suppress all of the details about other dimensions of performance except for those relevant to the analysis of performance along the dimensions of interest.

When the choice of a high-level technological solution defines sub-structures within which different technologies may be applied, the evaluation of the choice of technology at the highest level involves a special kind of abstraction in which the performance of several lower level objects is "aggregated" into a single higher-level performance measure. This is referred to as aggregation (Smith and Smith, 1977a). When a problem is decomposed in such a manner that each sub-problem is concerned with performing a limited set of functions, the process of aggregation is considerably simplified. Typically, this also leads to greater stability in the design process. That is, changes can be made to one aspect without substantially altering others. The impact of changes is localized. This feature can contribute to more efficient design implementations since more assumptions about the impact of changes on overall performance requirements can be more efficiently and effectively evaluated. In this way the task of design composition is simplified, and both the time to develop new designs and the effectiveness of the eventual solution may be enhanced.

The ease with which changes can be made in a modular design and its impact on overall performance evaluated is an important source of learning in the design process. Feedback is rapid, and cause-and-effect relationships can be more effectively evaluated as a basis for improving the level of understanding. As the general level of knowledge increases, it can be more easily applied to different design problems.

Another type of abstraction that has an important impact on the design process is referred to as generalization (Smith and Smith, 1977b). Generalizations are abstractions that enable a class of individual objects or artifacts to be thought of generically as a single object or artifact. It is an important mechanism in conceptualizing the real world and is a valuable means of learning. Children move from the observation of specific dogs to a model of dogs in general. Generalizations allow us to make predictions about the future on the basis of specific events in the past—if this fire and that fire have burned my hand, then perhaps fires in general will burn my hand.

The ability to generalize is an important means of problem-solving in design. Knowledge developed in the solution of one design sub-problem can often be re-used analogically in solving related sub-problems in the same design and for solving similar problems in other designs.

Examples of the use of generalization or analogical problem-solving in design include the re-use of modules or components of an existing design in creating related designs; the modification and re-use of information relating to the solution of one sub-problem in solving a related problem, and the use of knowledge gained in the problem-solving process (in the form of human expertise, heuristic procedures, or algorithms). The first two of these levels are particularly useful in addressing the "spatial" requirements, while the third is useful in addressing the dynamic requirements in manufacturing.

In general, therefore, modularity seeks to reduce development time by reducing the cycle time for each phase of the design and development process, to eliminate iteration between stages, as in the case of DFA and DFM, and to make it possible to transfer knowledge to related design projects.

MODULARITY—AN EXAMPLE

To provide an explicit illustration of the concepts discussed in the preceding section, we analyze an example from the microcomputer industry, which is one of the fastest-growing and most dynamic segments of the overall computer hardware industry. Several factors interact to shape the nature of competition in the industry. Among them are the rapid change in the underlying technologies, the existence of a global

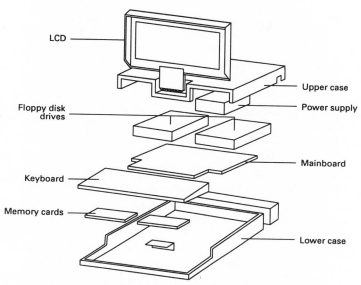

Figure 6.1. Assembly diagram for the PC convertible.

market for the products, and a large and growing number of international competitors.

To compete effectively in this industry, firms have had to acquire the ability rapidly to develop and introduce new products based on the latest technological advances. In addition, to protect existing markets from encroaching competition, the leading competitors have typically followed a strategy of producing products designed for a range of applications (Shirley, 1987; Ahlberg and McPhee, 1986). The ability to do this has been facilitated by the modular design of the microcomputers, the primary components of which are shown in Figure 6.1 (Newell and Bell, 1971). The modular structure of the product makes it possible to change one component with little if any effect on the other modules. With the emergence of international suppliers, each of whom produce a wide range of the major components, manufacturers of microprocessors can literally choose from extended libraries of modules/components and combine them into a system that meets the performance needs of their customers.

Because the gap in technological competence between the major competitors is small and declining, there is a pattern of new technological developments being quickly reverse-engineered and copied. While patents are an important means of protecting intellectual property in this environment, the pace of technological change has served to reduce the effective life of new technological developments (Ahlberg and McPhee, 1986). Since the advantages provided by new technological developments are as short lived as they are in this industry, it is important that firms develop the ability to produce products cost-effectively and to offer short lead times. With a wide and changing mix of products, the ability to achieve a high degree of flexibility in the production system is vitally important.

This brief analysis of the trends in the microcomputer industry highlights the rationale for the increased interest that has been demonstrated by manufacturers in the industry in designing products for ease of manufacturing and assembly. The dynamic nature of the industry also implies that it is necessary to design production systems so that they can not be used only to produce a wide range of products, but also so that they can be straightforwardly modified to accommodate a changing product mix.

In the remainder of this section, we focus on the design of a production system developed by a leading manufacturer of microcomputers for production of a new family of personal computers. There are several reasons for focusing on this manufacturing system. First, the manufacturing system is itself viewed by the company as a commercial product, which can be readily customized for use in the assembly of a range of other electronic products. The modular design of the production system contributes to the ease with which it can be customized for new applications. By analyzing the design of the system, we are able to illustrate some of the features of modular systems and their use in managing design constraints in practice. By discussing the concept in this context, the goal is to underline its generality. Possibly the most important reason for choosing this example, however, is that by analyzing the design of the production system, we are able to better comprehend the nature of manufacturing costs and the economics of design for manufacturing/assembly in these automated production systems. This is treated more fully in the following section.

The production system with which we are here concerned was developed in parallel with the design of the new products, and a key design objective was that all activities involved in the production of the microcomputer were to be performed in a manually untended manner. These activities were to include the receiving and storage of raw materials, transportation of the materials to assembly cells, assembly of the products, test, burn-in, inspection, and packaging of the final product. Major redesigns of the family of computers were expected to take place every eighteen months, and the system was to be flexible enough to accommodate changes in the physical composition and size of the product as a result of these redesigns. Because of the competitive nature of the market segment for which the products were being designed, tight constraints were placed on the overall cost of the production system. Finally, because the firm produced a complete line of computers and peripheral devices and was also a producer of automated assembly systems for sale to other electronic and electromechanic manufacturers, an important goal in developing the system was to create a design that could be easily customized for use in other internal production operations as well in the commercial systems. These objectives established performance constraints within which the design engineers were to work. The "space" of potential design solutions was, however, very large.

The time available for development was constrained by the need to have the products available to meet the narrow window of opportunity that existed for the firm before the technologies were made obsolete by advances in semiconductor technologies and before similar products were offered by any of the other large and aggressive group of international competitors. Because of this, the systems had to be designed using currently available technologies. These conditions imposed tight realizability constraints on the design project.

The limited space available for installation of the final system and the desire to develop a compact system that would be more marketable to outside customers established loose compositional constraints of a mechanical nature on the design. Com-

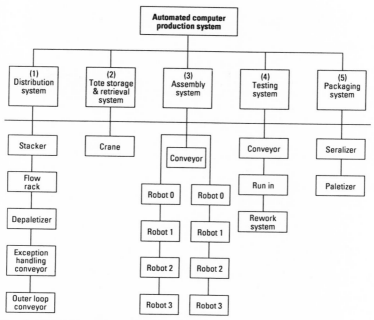

Figure 6.2. Automated computer production system.

positional constraints on the electronic aspects of the design were imposed by the desire to have an integrated yet easily maintained and modifiable control system for the production facility.

The complexity of the project is evident from this brief but incomplete description of the initial goals. It is important to note that even with a clear statement of the design objectives, there remained a number of sources of potential ambiguity in the project. Because the system was to be developed in parallel with the product, many of the final specifications of the product were unavailable at the outset.

To deal with this complexity, a decision was made to adopt a modular approach to the design of the system. The five major modules into which the system was decomposed are shown in Figure 6.2. Each of these was in turn decomposed hierarchically into several sub-modules as illustrated. By decomposing the design of each of the major modules into a series of progressively smaller but more well-defined problems, the details of the design at low levels of the hierarchy did not have to be made until that level was reached in the process. In this approach, high-level design decisions relating to the major modules could be made even before final specifications of the product were developed. In the case of the material-handling system, maximum component dimensions were established and these were used to establish the dimensions of the totes to be used to transport materials throughout the system. Note that the establishment of maximum product dimensions served to establish binding compositional constraints for the designers of the family of microprocessors that was being developed in parallel. This in turn was used to determine the dimensions of the pallets in which raw materials were to be delivered to the system. The combined effect of having made these decisions was to make it possible to proceed with

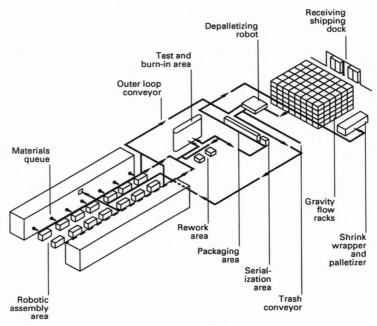

Figure 6.3. Schematic of the automated assembly process.

the mechanical design of the flow rack, the depalletizing unit, the tote storage and retrieval system, the outerloop conveyor, and the outerloop conveyor in parallel with each other and each in a fairly independent manner. A brief description of each of the major modules and the sub-modules of which they are composed is presented in the Appendix. A schematic diagram of the system is presented in Figure 6.3.

Each sub-module is composed of at least one electromechanical device such as a robot, control of which is effected using microcomputers. By design, each sub-module could be operated independently of the other sub-modules. To integrate the execution of the production activities at the different sub-modules, another microprocessor, referred to as a supervisory computer, was used at the module level. This provided timing instructions to each of the sub-module controllers and passed information on the status of activities to a minicomputer that was used to coordinate the activities of the five major modules.

Among the important advantages of the modular approach to design in this instance were the following:

- Each of the modules could be developed in relative isolation from the others. This allowed for greater parallelism in the product development process, and reduced development time for the system.
- When modifications are necessary in any of the modules, it is comparatively easy to isolate a module from the remainder of the system, and the changes are typically localized to a single module. This considerably simplifies the process of making changes, for example, to accommodate the assembly of new classes of products.
- Documentation is considerably simplified.

- Problem solving is simplified, as errors are often confined to a single system and are comparatively easy to trace.
- By using standard interfaces and protocols for communication among the controllers in the different (sub)-modules, changes to the composition and configuration of the system could be easily accommodated. Thus, for example, new assembly sub-modules that incorporated more advanced robots could be introduced as they became available and the cells redesigned to include different numbers of machines as the need arose.

The ability to incorporate a number of different assembly cells or modules, each of which is supported in common by the functions performed by the remaining modules, allows the system to be used to produce several families of products simultaneously, and to utilize all of the equipment productively. Given that most of the production costs (excluding materials) are fixed and sunk before the first unit is produced, the effective utilization of the facility is an important determinant of the cost of production on a per unit basis. In addition, the ability incrementally to adapt the system for manufacturing different classes of product by reprogramming and reconfiguring one or more of the assembly cells allows the system to be adapted for use in manufacturing a changing mix of products.

The architecture of the system is quite general. The design of the modules and the versatility of the robots and the ability to vary the number of these robots in a cell or module allow systems such as the one described above to be used for a wide range of applications. In developing subsequent assembly systems, it is possible to re-use modules of the existing system ''as is,'' or to modify the designs to create variants. This is illustrated graphically in Figure 6.4. Where documentation of design decisions is complete, the time and cost involved in modifying the designs for changes in the products to be assembled or for use of the module in different systems are comparatively small. In the most general form, the knowledge generated in the design of this system can be re-used in developing systems for use in other applications. Many of the major modules and sub-modules of the system described above have been used in the assembly of other families of computers, printers, typewriters, as well as various types of automotive electronic components.

THE ECONOMICS OF MODULAR SYSTEMS

There are, in general, three classes of cost with which design and production managers must be concerned in the development and operation of systems such as the one described above. The first of these are variable costs, which can be uniquely identified with each item of production. These are sometimes referred to as ''unique'' costs. Conventionally, the main unique costs in production systems have been the cost of materials and direct labor used in producing each item. In systems such as the one we have described, most of the tasks traditionally performed by direct labor are performed by robots, the costs of which are fixed.

The second category of costs is referred to as ''joint'' costs, these are costs incurred as a result of the decision to produce a family of related products. The joint costs of production include the costs of tools and fixtures that are shared by a family

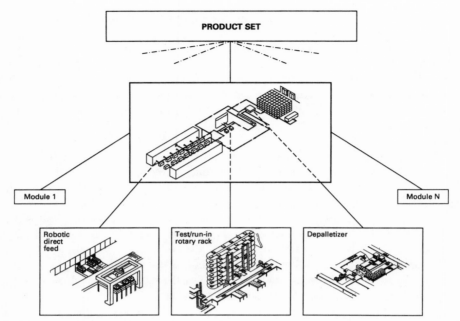

Figure 6.4. Product set.

of products. If a particular cell is dedicated to the production of a family of products, these are also joint costs of production shared by the products that make up the family.

The final category of costs is referred to as "common" costs. Common costs include the costs of those modules shared by all products made in the system. For example, the cost of developing the distribution system, the tote storage and retrieval system (TSRS), and the testing and packaging modules is shared by all products produced in the system and is therefore a common cost.

A primary goal of most design and manufacturing managers is to allocate resources to the development and production of products so as to maximize the contribution obtained from operating their systems. Normally, resource allocations are subject to constraints on availability. The general resource allocation problem therefore can be written as follows:

$$\text{Maximize } P = \Sigma \, [(P_u - C_u) \, X - C_j \, Y - C_c \, Z] \tag{1}$$

Subject to:

$$\Sigma \, [A_u \, X + A_j \, Y] \leq B \tag{2}$$

$$\Sigma \, [D_j \, Y + D_c \, Z] \leq B' \tag{3}$$

$$X \geq 0, \ Y, \ Z \geq 0 \text{ and integer} \tag{4}$$

where P_u, C_u, and X are the revenue, unique costs, and production quantities of a specific product, Y is a $(0,1)$ variable that is positive if a decision is made to produce a family of products, and C_j is the joint cost associated with developing this family of products. The D_j coefficients are the resources consumed in the development process that are "jointly shared" by members of the family of products. Z is a (0,1) variable denoting the decision to develop a specified set of common resources. C_c are the common costs of developing these resources, and D_c are the common development and manufacturing costs. B' are those design, engineering, and capital resources incurred as a result of the decision to develop joint or common resources only.

Equation 1 is the objective function that specifies the objective as the maximization of contribution from operating the system. Constraints 2 are the production constraints; A_u X is the setup cost associated with producing the product X; and A_j Y is the setup cost associated with producing family Y. Constraints 3 are the development constraints, and constraints 4 are the non-negativity constraints and specify Y and Z to be integer variables.

The problem as represented is a mixed integer program, and can accommodate a variety of relationships between the x and y variables and between the y and z variables. This general class of problems can be solved using the Bender's decomposition technique (Benders, 1962).

In conventional manufacturing systems, the utilization of direct labor has typically been a primary concern, and with direct costs dominating the fixed costs of production, a normal practice has been to combine the common and joint costs of production and allocate these on the basis of direct materials or labor costs to the individual products. The problem decomposes into a design problem, where the objective is to minimize the cost of development subject to constraints 3, and a planning problem, where the goal is to maximize contribution to operating profit (the first component in Equation 1) subject to constraints 2. Much of the work on design for manufacturing has focused on reducing the C_u and A_u coefficients.

In automated production systems such as the one we have described, however, the A_u coefficients are established in the design process, P_u is exogenously determined in most instances, being set by market forces, and C_u consists primarily of material costs in systems such as the one we have described above. In this environment, then, the primary decisions over which the production manager has control in order to improve performance are those associated with the choice of families and common resources to develop and the design of the production system. Modularity focuses on reducing the D_j and D_c variables and provides a mechanism by which the manager can utilize the design and development resources more effectively in developing a range of products belonging to different families.

An important aspect of the formulation is that it emphasizes the relationship between the various types of costs incurred in the system. It illustrates that in automated production systems such as the one described above, it is not meaningful to ask the question, What does a product cost? A product is made because of its fit with the existing set of assets and its ability to improve the contribution from operating the system. Here the assets include the fixed resources as well as the store of information on the design of related products. Products that fit with the existing portfolio incur little additional fixed costs in production. Modularity, by reducing the cost to develop "variants" of existing families and reducing the cost of development of new families, is therefore a potentially important tool in improving performance in both the design and manufacturing processes.

CONCLUSIONS

A primary theme of this chapter has been that in order to comprehend fully the nature of the task involved in designing products for ease of manufacturing and assembly, it is first necessary to understand the nature of the manufacturing task imposed by recent changes in the nature of industrial competition and technology development. By analyzing the current environment faced by manufacturing firms in a wide cross section of industries, it was established that in addition to minimizing fabrication costs, firms are increasingly concerned with making a wide range of related products and pursing the development of a changing mix of products over time. Typically, research on design for manufacturing has focused on the problem of reducing production costs. By analyzing the process of design in some detail, it was shown that the conventional methods have sought to achieve these objectives by imposing tighter constraints on the design process. Several types of constraints were analyzed. The goal of this chapter has been to present a complementary approach that also addresses the issues of developing and producing a large and changing mix of products. The concept of modularity in design was presented as such a method. The implications of this method on the design, development, and production processes were analyzed and the impact on the economics of production discussed.

The intent of this work has been to focus attention on the structure of the design task, and to begin to model the utilization of resources in the product development process. Areas in which additional work currently is focused include the development of the specifications for a system that allows for the capture of important aspects of design knowledge within an organization (Shirley and Eastman, 1989). This system provides a formal method of recording and archiving information on the sequence of design decisions made over the course of a project so that this information may be more efficiently revised and re-used in the development of related projects.

REFERENCES

Ahlberg, R., and J. McPhee. *A Competitive Assessment of the U.S. Microcomputer Industry: Business/Professional Systems.* U.S. Department of Commerce, 1986.

Benders, J. "Partitioning Procedures for Solving Mixed-Variable Programming Problems." *Numerische Mathematik* 4, 1962, 238–252.

Boothroyd, G. "Design for Assembly—The Key to Design for Manufacturing." *International Journal of Advanced Manufacturing Technology* 2(3), 1987, 3–11.

——— and P. Dewhurst. *Product Design for Assembly.* Wakefield, R.I.: Boothroyd Dewhurst, Inc., 1987.

Bracken, F. "Parts Classification and Gripper Design for Automatic Handling and Assembly." *5th International Conference on Assembly Automation,* 1984, 181–189.

——— and G. E. Isolia. "Design of Data Processing Equipment for Automated Assembly." In *Programmable Assembly,* ed. W. Heginbotham. Kempston, Bedford, England: IFS Publications, 1984, 105–126.

Clark, K. B. "Managing Technology in International Competition: The Case of Product Development in Response to Foreign Entry." *Proceedings of the NBC/KEG Conference on International Competition,* 1985.

——— and T. Fujimoto. "Product Development and Competitiveness." *Proceedings of the OECD International Seminar in Science, Technology and Economic Growth,* Paris, 1989.

Eastman, C. "Automatic Composition of Design." In *Design Theory,* eds. S. Newsome et al. New York: Springer-Verlag, 1988, 158–172.

Eaton, B., and R. Lipsey. "Introduction of Space into Neo-classical Model of Industrial Economics." In *Studies in Modern Economics,* eds. M. Artis and A. Norbay. Oxford: Basil Blackwell, 1976, 59–96.

Lancaster, K. *Variety, Equity and Efficiency.* New York: University Press, 1979.

Lee, D., and B. Hight. "Design for Assembly: A Bibliography." UCLA Manufacturing Engineering Program Working Paper, 1989.

Marple, D. "The Decisions of Engineering Design." *IEEE Transactions on Engineering Management,* 1961, 55–71.

Newell, A., and C. Bell. *Computer Structures: Readings and Examples.* New York: McGraw Hill Book Company, 1971.

Parnas, D. "Information Distribution Aspects of Design Methodology." *Proceedings of the 1971 IFIP Conference,* 1971.

————. "A Technique for Software Module Specification with Examples." *Communication of the ACM* 15(5), 1976, 330–336.

Poli, C., R. Graves, and R. Groppetti. "Rating Products for Ease of Assembly." *Machine Design* 7, 1986, 79–84.

Redford, A. H. "Product Design for Automated Assembly." *Assembly Automation* 8, 1984, 133–136.

Schmalensee, R. "Entry Deterrence in the Ready to Eat Breakfast Cereal Industry." *Bell Journal of Economics* 9, 1978, 305–327.

Shirley, G. "The Management of Manufacturing Flexibility: Studies in the Design/Manufacturing Interface." Harvard Business School Doctoral Dissertation, 1987.

———— and C. Eastman. "Specification of a Knowledge Based System for Product Design and Engineering." UCLA Working Paper, 1990.

Simon, H. A. *The New Science of Management Decision.* Englewood Cliffs, N.J.: Prentice-Hall, 1960, 112.

Smith, J., and D. Smith. "Database Abstractions: Aggregation and Generalization." *ACM Transactions on Database Systems* 2(2), 1977a, 105–133.

———— and D. Smith. "Database Abstractions: Aggregations." *Communications of the ACM* 20(6), 1977b, 405–413.

Stevens, W. P., G. J. Myers, and L. Constantine. "Structured Design." *IBM Systems Journal* 13(2), 1974, 115–139.

APPENDIX

In this appendix we present a brief description of the function performed by each of the modules and the way in which they are integrated.

The Distribution System

Distribution is composed of five sub-modules. In the first of these, incoming raw material pallets from suppliers are staged by a stacking crane and placed in an automated flow rack. Material pallets are constructed to accommodate a specific number of specially designed cardboard containers in which components are packaged either individually or in matrices. The containers are designed to retain the components in an orientation that is immediately accessible to assembly robots.

From the flow rack, pallets are fed to a depalletizing robot as needed by the process. This robot removes the containers, the common footprint dimensions of which allow them to fit neatly into standard component totes that are used to transport them throughout the system. The tote with components is placed by the robot onto the distribution conveyor from which it is transferred to the outer loop conveyor. Before proceeding, however, the barcode on the container is read and the tote weighed and sized for uniformity. Improperly loaded containers are diverted to an exception handling spur on the distribution conveyor.

As its name implies, the outer loop conveyor is designed to allow totes to circulate in the event that they cannot be immediately off-loaded to the tote storage and removal system (TSRS). The outer loop conveyor is also equipped with a series of accumulation zones for queuing of the totes at various control points along its path.

Tote Storage and Retrieval System

On entering the TSRS the tote is again scanned before being stored by the crane, which alternates between two multitiered storage racks. The TSRS crane has the ability to deliver any tote from any storage location to any of a variable number of exit conveyors along its length. These exit conveyors are used to feed the component totes into the robot work envelope, and are controlled by the TSRS workstation controller. When components are required by the assembly workstation, the latter removes a tote filled with components from the exit conveyor delivery point. On sensing a missing tote, the TSRS controller issues an instruction to the TSRS crane to have it replenished. This arrangement effectively decouples the operation of the TSRS from the robot cell, allowing both to operate as stand-alone modules.

Robotic Assembly System

The robot assembly line is composed of two identical assembly cells containing three robots each. These intelligent cartesian robots are cantilever mounted and are equipped with vision and tactile sensing capabilities, and each has seven degrees of freedom. Pallets with base plates are fed to the two cells by a pair of pick-and-place robots via a constantly powered roller conveyor. At the robot workstation, the pallet is diverted off-line into a one-pallet buffer zone, before being transferred into the robots work envelope, where the assembly operations occur. This arrangement effectively creates an asynchronous assembly system. This is important in absorbing the small imbalances that might arise when it is necessary to invoke an error recovery routine during the process of performing a particular operation at one of the stations. These routines allow the robot to repeat steps when a contingency is observed, or to branch to a completely different sequence if required. For example, if in the process of installing a part, the robot senses that it is slightly misoriented, it can branch to a subroutine to place the part into a locating fixture, regrasp the part, and continue with the assembly operation.

In the assembly process each robot, under the control of its station controller, performs a specified sequence of operations before passing the pallet to the conveyor for delivery to the adjacent robot station. In producing the personal computer, 20 components are assembled in a process that requires 26 discrete operations. This number includes some subassemblies such as the keyboard and LCD (liquid crystal diode) but excludes fasteners. Tasks include joining two memory cards into a single unit and inserting them into the lower case, mounting floppy disk drives, attaching the keyboard and LCD, and mounting and fastening the power supply in place. An illustrative sketch of the assembly is presented as shown in Figure 6.1.

Test System

Completed assemblies are passed from the assembly via a conveyor to the test and run-in area. Here, after the barcode has been read, the test media is inserted into the disk drives by a robot and transferred to a multitiered test rack where "burn-in" occurs. In the rack, the unit is subjected to a battery of tests, some of which, such as testing the keyboard, adjusting the display, and visual confirmation of the results, are performed by another robot. Defective units are passed to a rework area by a conveyor loop, where they are manually repaired. In resolving the error, the test media is read by a computer that determines the source of contingency and the appropriate procedure to follow for correction. The required sequence of tasks is displayed for the operator on an adjacent terminal. Once repaired, the machine is returned to the rack for further testing.

Packaging System

Products that successfully complete the process are serialized by laser, removed from the build pallet by a robot, packaged, stretch-wrapped, and staged to await shipping. All of these tasks are performed automatically under microprocessor control.

The Control Architecture

The control architecture of the system is a three-tiered hierarchy. At the top of the hierarchy is a single high-level computer module (the area controller). Here, most global goals are decided upon. Once these decisions are made, the entire hierarchical structure is committed to a unified and coordinated course of action to achieve these goals. Below this in the hierarchy are five area controllers—the distribution system controller, the TSRS controller, the assembly cell controllers, the test area controller, and the packaging area controller. The area or cell controllers schedule jobs and route totes and workpieces through the cell/area. They also make sure that each robot/ machine has the proper tools at the proper time to perform the required work on each part. Each cell controller supervises several workstation control units. These include the individual robots, smart sensors, conveyors, and the tote storage and removal system. Each of these in turn may have its own internal hierarchical control system.

II

DFM AND THE NEW PRODUCT DEVELOPMENT PROCESS

7

CONCEPT DEVELOPMENT EFFORT IN MANUFACTURING

JOHN E. ETTLIE

The resurgence of intense interest in managing the product development process in manufacturing is but one of many manifestations of the increasing harshness of global competitiveness. New and improved products were once considered to be central to the competitive strategy of firms in just a few, less mature industries. New products are costly to develop and risky to introduce (Wind and Mahajan, 1988). Now all manufacturing industries, as well as many segments of the service, extraction, and agriculture industries seem to be preoccupied with the hows and wherefores of getting to market sooner with products that please customers. But this does not remove the risk or expense. For example, Grabowski and Vernon (1990) found that only the top 30 drugs introduced in the recent history of the pharmaceutical industry actually made enough money to cover mean R&D costs.

The focus of this chapter is on the early stage of the product development process called concept development. It is defined as that stage of the innovation process when customer needs are converted into ideas for new outputs—like new products, new processes, or new services, or a combination of all three, which often happens. There are assumptions embedded within this definition that will become more obvious when it is compared to the other definitions offered later in the chapter.

CONCEPT DEVELOPMENT IS UNIQUE

Concept development is a unique stage in the innovation process that has not received the attention that it should, given its importance. With all the attention being given to speeding up the time to market with new products, it hardly pays if the product specification is wrong, and the early stages of concept development often determine what these standards will be.

There are at least four things that make the concept development stage of the innovation process unique. First, it is the only obvious stage of the process where individual factors meet group process factors. Jelinek and Schoonhoven (1990) present a rather detailed picture of the early stage of the innovation process in their research on Silicon Valley firms. They call the early-stage filters "encouragement as well as choice'' (p. 177). They argue that this is not so much a stage in the innovation process as an ongoing activity of continuous screening—both by individuals of their own ideas and by groups. People decide what to work on, and by default, this aggregate is an organizational choice. Time is the scarce resource and researchers are typically 110% busy. Before ideas are born to the world, they must be extraordinary in concept—both technologically and commercially. Getting other people to "buy in" to the idea and shift from personal to others' resources is another critical step. This two-step process of "self-selection" and "buy in" by others, operates together, rather than as wholly discrete stages (p. 178). This is the place where the individual meets the group, and it is essential to understand this as part of the concept development process. When group creativity takes over, we normally think of the development process beginning (Hage, 1980), but it clearly is more subtle than that and this is what makes the concept development stage unique. What is worthy and what is not, what is filtered and what is not, is all part of this process.

A second characteristic of concept development that makes it unique comes from understanding concurrent, simultaneous, or overlapping engineering of product and process (Clark, Chew, and Fujimoto, Chapter 11, this volume). As Eppinger and colleagues (1990, p. 15) have aptly pointed out, "Not everything can be 'simultaneous' in simultaneous engineering." The issue of how to coordinate effort and thinking but not involve everyone all the time and inappropriately is a real challenge in a world that is convinced of the virtues of concurrency. But if the overlapping engineering charts in the work by Clark, Chew, and Fujimoto are examined carefully, one sees that the early stages—qua concept development process, i.e., concept generation, product planning, and advanced engineering—overlap very little with product and process engineering. This part of development stands alone as a unique and, perhaps, less well understood part of the overall process. Nevertheless, project success still depends on closely coordinating the concept development stage with the stages that follow it.

The third feature of concept development, the way in which this process proceeds, is problematic. Pugh (1981) argues that the reasons designs fail are due in part to the lack of thoroughness in the "conceptual approach" that is to be established early in the design cycle. Worse, in a way, is finding the right design but not knowing why it is the best approach. Pugh observes that part of the problem stems from the fact that experienced designers want to "get on with it" and resist using any formal design methodology or procedure that might enhance thoroughness in concept development. Extending this point, Welsch, Dehler, and Green (1990) show that the early concept development stage, when initial commitment and project priority are established, is highly contingent upon the interaction of strategic objectives and technological stimulus in R&D. They found for a sample of 80 R&D projects that not only can these projects be correctly classified according to their strategic objectives, but there is also a significant interaction between technical stimulus and the strategic objectives of projects. For example, projects with low technical stimulus that were strategically defensive had the lowest initial management commitment. Somers and Birnbaum (1990) argue that in high-technology firms strategic alliances are a primary consideration in product and process development. Clearly, the concept development stage of projects is where strategy and stimulus intersect.

Automobile industry examples further illustrate this unique aspect of concept

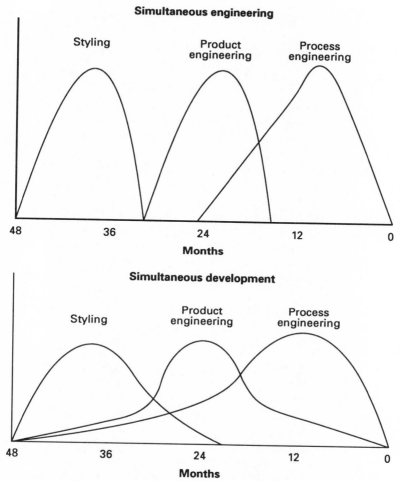

Figure 7.1. Simultaneous engineering versus simultaneous development of new automobile bodies. Source: Smith, 1990, pp. 5, 7.

development. In a study of automotive body development, Smith (1990) reports that of the three alternative approaches to design, simultaneous development (rather than sequential or simultaneous or concurrent engineering) produced the best results, saving as much as $300 to $400 per body. Simultaneous development involves the overlapping, democratic participation of styling with product and process engineering. The latter groups typically form simultaneous engineering groups, and styling proceeds independently in typical body design practice. Although styling exercises the greatest authority over designs, representatives of this group rarely deal directly with manufacturing engineers. These two approaches are contrasted in Figure 7.1 from Smith (1990, pp. 5,7).

Not surprisingly, General Motors Corporation has drastically altered its new product development process to take into account findings from these types of studies. At a recent symposium, Hoglund (1990) outlined this new design practice methodology at GM that is called the "product program management process." Its distinctive feature, for this discussion, is that it devotes considerable effort on the first (zero) phase of the process. This early phase involves concept initiation, product proposal development; concept direction, definition, and development; and concept approval. Hoglund (1990) reported that the real challenge to making this all work is to have general managers stay out of the development cycle once significant approval milestones have been passed. GM has not always been successful at doing this, and GM is probably very typical of most manufacturing firms.

Fourth, and finally, idea generation and creativity in organizational settings continue to be rather perplexing aspects of organizational life. Little work has appeared since the article by Ettlie and O'Keefe (1982) on occupation-independent measures of innovativeness for individuals. Since the concept development stage involves the interaction of individual self-screening and group filters, more needs to be known about the role of creativity in this stage of the process.

In summary, concept development is unique because it represents the intersection of individual and group processes, it is a stage of the product development process often excluded from concurrent engineering, and it combines elements of strategy and technological stimulus. Lastly, creativity that is occupation independent is still not well understood.

A CASE HISTORY: THE MAZDA MIATA

It might be useful to explore an illustrative case of a successful new product to help understand why concept development and effort expended at this stage of the innovation process are crucial to success. The story of the Mazda Miata was recently published by Levin (1990) and serves as an instructive example for this purpose.

The Miata (code-named the P729 in 1983 when the feasibility study was approved) was conceived by Bob Hall, formerly an automotive journalist for *Motor Trend,* who suggested that Kenichi Yamamoto, then managing director of R&D for Mazda, but destined to become chairman, take a ride in a Triumph Spitfire. Hall believed in the small British sports car like the Spitfire and was saddened to see it leave the auto scene—an exit that was well in progress by 1978. Yamamoto took the ride and met Hall in 1979. Hall then sketched the idea for a small Mazda sports car on the blackboard. Hall was hired in 1981 as a product planner at Mazda's Irvine California studio but turned to working on the van project.

When the sports car project was approved in 1983, two teams—one in Japan and one in California—started parallel development efforts. This was a risky project for Mazda or any other car company, because a lightweight, relatively inexpensive sports car was a novel idea. The car would not be fancy, so it had to be inherently appealing (it winks at you in the showroom—as some put it). The Japanese were working on two concepts—front-wheel-drive/front engine and mid-engine/rear-wheel drive. The American team was working on a front-engine/rear-wheel drive design, primarily because of the handling characteristics of this approach. They wanted to avoid the "wheel hop and tire spinning at low speeds in tight corners," characteristic of front wheel drive cars (Levin, 1990, p. 75). The panel of Mazda executives chose the American design (by Bob Hall, Mark Jordan, and Tom Matano) in its convertible incarnation in 1984. It took eight months to make a prototype, which was completed in the fall of 1985. In October 1985 the car was test driven from Los Angeles to Santa Barbara and attracted so much attention that the Mazda board of managing directors approved production of the car without focus group and market research tests.

The production budget was spartan. Capital allotted to the project was $100 million to $200 million. A line could not be built for that amount so the Miata was scheduled for a flexible assembly line already producing the 929 and the 626 in Hiroshima and Hofu. The car was formally introduced in January 1989, and 23,000 units shipped to the United States were sold out in the first six months. At first, red, white, and blue were the only colors available. In 1990, silver was added. The Miata uses lightweight, high-strength metal panels and aluminum components to save weight, so that it tips the scales at 2,182 pounds or 150 pounds more than an MGB. It gets 25 mpg city and 30 mpg highway. Consumer testing agents have turned up no major flaws in the car.

If one uses approval of concept (1984) as the start of this project, this was a four- to five-year effort for Mazda. That's not remarkable but it is not slow. More importantly, the idea for the Miata was 12 years old when it was introduced—a relatively long incubation period. It was risky and used unorthodox market test methods. It resulted from the parallel effort of teams in competition. Cost to produce was crucial, and flexible production and assembly were needed. No radical technology was proposed or incorporated based on the Levin (1990) account.

CONCEPT DEVELOPMENT FROM THE R&D AND MARKETING PERSPECTIVES

Although there are many articles in the literature that present ideas and conclusions relevant to concept development, most of this material is actually tangential to this stage of innovating. In particular, there is little attention in any of these articles to the issue of optimizing the concept development process. That is the focus of this chapter.

One reason for this state of affairs appears to be that most treatments of the subject are bound by single disciplines. Historically, either marketing or R&D has claimed ownership for this stage of the innovation process. The R&D management perspective has traditionally concentrated on creating a technological base in an organization that is managed as the well spring of new ideas. The design department or teams of product planners, which often include general managers, is typically the focal point where many ideas mature and are converted to tentative plans for manufacture.

The second school of thought, that appears to have developed quite independently, is the marketing perspective. In the extreme, this school rejects the idea of concept development altogether when there is any absence of market information: that is, there can be no concept development without systematic customer information (see Conway and McGuinness, 1986; Crawford, 1983, p. 279). Marketers do not insist that the marketing department be responsible for this information, but that some mechanism be in place for this intelligence to be incorporated into the concept development phase.

Idea management in marketing terms is a process of screening customer-dominated sources of information, and creativity is defined as generating innovative ideas to expand or open new markets (Urban and Hauser, 1987, p. 70ff.). Unsolicited ideas to satisfy customers are often rejected because of legal ownership issues and because ideas cannot be evaluated simultaneously when presented from the outside in this manner (Ottosson, 1983). Therefore, systematic, proactive information gathering and management are at the heart of this approach—especially for consumer markets.

Occasionally, authors have argued convincingly for a balanced perspective on this issue. For example, Souder (1987, 1989) has studied the R&D-marketing interface and concludes that some combination of technological capability and market knowledge is optimal. Mowery and Rosenberg (1979) show that "market-pull"-dominated theories of successful innovations are flawed in concept and empirical methodology. Zinger and Maidique (1990) found that R&D, synergy of the new product with existing competencies, and to a lesser extent marketing determined the success and failure of 330 new products in the electronics industry. What is more, even when this "blended" approach to product innovation is used, the creative part of the process—matching market needs with capabilities—looms as a very significant part of the effort. Ergo, the focus here on concept development and optimization of this early stage in the innovation process.

IDEA GENERATION

Perhaps the first systematic study on idea generation was done by Baker, Siegman, and Rubenstein (1967). Among other things, this early study focused on the issue of why some ideas never get submitted in R & D settings. A later, follow-up work by Baker, Green, and Bean (1985) is very relevant to understanding concept development. The authors collected data on 211 R&D projects done by 21 companies in four industries: steel, agricultural chemicals, food processing, and industrial chemicals. They define an idea as "a potential proposal for undertaking new technical work which will require commitment of significant organizational resources" (p. 35). Idea generation, on the other hand, has been variously defined as the "coming together of an organizational need, problem or opportunity with a means of satisfying the need, solving the problem, or capitalizing on the opportunity" (p. 35). Baker et al. (1985) were particularly interested in the role played by the individual(s) that articulate the performance gap that defines the need for a new idea.

The most important findings of the study are as follows:

1. The type of idea generated is related to the source of first suggestion under certain circumstances. For example, marketing (including distribution and sales) and the customer are more likely to be the source of new product ideas, whereas new process projects are more likely to be initiated by ideas from production, engineering, or technical services. Interestingly, technical sources inside or outside

the firm were not found to be related in a statistically significant way to project type, although they represent the largest number of ideas for new projects (142 of 726 ideas).

2. The source of ideas is related to success or failure of new product projects but not for new or modified processes and not for projects that combine new products and processes. In the case of new product projects, they are least likely to succeed when R&D is the sole source of the first idea, and most likely to succeed when marketing and/or the customer is involved in the first idea. The probability of success was 35% in the former and 65% in the latter. When source of ideas is confined to just new products and just a single source (e.g., R&D versus marketing) there was no difference in success rate. Yet there is a difference in the type of project that is the sole source in its initial idea. When R&D was the sole source, the project was less well defined, technically more complex, and riskier, as one might expect.

3. Source of first idea for projects varied by industry. Consumer-oriented industries like food tend to favor marketing for ideas, whereas industrial markets favored the customer directly (flat steel and chemicals) or R&D staff (agricultural chemicals).

Further data on why ideas are not submitted is also reviewed by Baker et al. (1985). Since performance of regular tasks in R&D often takes precedence over idea generation, it is not surprising that all new ideas are not submitted. However, another common reason for failure to submit ideas is that thoughts are not well developed or are incomplete at the early stages of this process. Managers tend not to reward incompleteness and therein lies a barrier to creativity.

Conway and McGuinness (1986) report some interesting findings on idea generation in a sample of nine Canadian technology-based firms. Regardless of product type, management and market considerations initiated the search process for new products rather than technological capability, which becomes a sustaining force later in the development process. This early stage of the innovation process is characterized by the authors as a very difficult information-processing problem. They describe it variously as ''concept generation . . . exploration . . . opportunity identification . . . and idea generation'' (p. 279). Typically, firms tend to follow procedures that have worked in the past when searching for new product ideas.

Conway and McGuinness (1986) reviewed the development of 35 product concepts in these nine firms. Although these firms had well-established methods of finding concepts for new products, few formalized these methods. Approaches tended to fall into one of two categories: the first relies heavily on defining and communicating corporate purpose and strategic value, and the second relies on a dominant mode of search based on past experience. For example, six of the nine firms had a specific market-driven corporate development strategy. Customers and markets are primary sources of information for this approach, as would be expected. But interpretation of market data is not easy because one has to find the ''common denominator'' that defines a new product.

Bar (1989) proposes a systematic technique for new product idea generation that involves using the Dialog databank and search for root concepts and words relating to ideas that are innovative. The technique is demonstrated for cases: new uses for water and new uses for metal-powder technology. Problems with the technique are that it tends to generate large amounts of irrelevant information and that it cannot provide truly novel ideas.

PRODUCT DEVELOPMENT

In a recent review of the literature on new product performance, Lilien and Yoon (1989) review 17 empirical studies on industrial innovation. Only two of these empirical studies suggest that the concept development stage of the process is directly related to the ultimate outcomes of the innovation process. Constandse (1971) suggests that absence of "people-orientedness in idea generation" is related to product failure, and Cooper (1986) concentrates on new product screening as the missing link in the development process that could increase the hit rate of new product ideas. Ram and Ram (1989) recently published an expert system to aid in this new product screening process and use the example of the financial services industry in their research. They argue for this choice of industry for their demonstration because this industry has experienced "explosive growth" in the last ten years with a "tremendous increase" in the number and types of new services offered.

Lilien and Yoon (1989) go on to test their ideas derived from this literature review using a database of 112 new industrial products from 52 French firms. They found that original new products and reformulated new products do not follow the same pattern of factors that determine successful introduction. For original new products, greater production and marketing expertise are required for success, market growth rates are higher, and competition intensifies when the time lag to market is longer. For reformulated new products, production expertise and competition are success factors but not marketing expertise or growth. Other research has also shown that radical versus incremental new products and processes require different mechanisms for successful introduction (Ettlie, Bridges, and O'Keefe, 1984; Nord and Tucker, 1987). Schoonhoven and co-workers (1990) found for new ventures in the semiconductor industry that substantial technical development slows time of introduction of first products. For faster introduction, they recommend that "less technically ambitious projects should be selected" (p. 201). Ettlie (1983) found that government-sponsored incremental technology innovations were more likely to be commercialized.

In addition to type of new product (original versus reformulated), the stage of the development process appears to be critical in understanding success factors. Pinto and Slevin (1989) have reported on a study of 159 R&D projects in several industries and found that the critical success factors vary at each stage of the project. During stage 1, the conceptual stage, four factors, including a factor that did appear in the other three stages as well—project mission (goal clarity)—were most important. Having initial clarity of goals and general direction, accounted for 67% of the variance in project success. Client consultation, quality of personnel, and urgency of the project accounted for another 25% of the cumulative variance in project success. Later in the life of these projects, other factors show up, for example, environmental events, top management support, and technical task accomplishment.

DESIGN-MANUFACTURING INTEGRATION

A number of studies are beginning to appear that focus on the management of the design process in a way that anticipates the impact of down-stream decisions made early in the process. For example, Salzman (1990) reports on a study of 63 manufacturing firms in the metalworking industries (e.g., air conditioners, computers, etc.) focusing on "skill-based design" or the involvement of the work force in design decisions. Most companies did not have an explicit policy "specifying worker participation in the design process" (p. 2), although 32% of these firms do involve workers informally. Only 8% required worker participation in equipment selection or on committees. Case evidence suggests that when workers are involved in new system design, firms experience more success with the program.

Trygg (1990) reports on three case studies in Swedish manufacturing with particular attention to how manufacturing and design are coordinated for design purposes. He measured location of departments (co-location versus distance to communicate face-to-face) and other factors. The impact on engineering change frequency was the object of the study. An important finding of this study was the size dependency of these relationships: "The larger the organization, and hence more complex, the greater the need for effective coordination, which consequently leads to greater use of integration mechanisms" (p. 7). He calls for better measurement of integration methods. Rothwell and Whiston (1989) also propose that size and structure interact with integrating mechanism choice.

In a study that attempted to use reliable measures of design-manufacturing integration, Ettlie (1990) reported significantly higher new system utilization when greater integration was accomplished. Integration was measured by five three-point factors: (1) outside training for the design team in design-for-manufacturing techniques, (2) manufacturing sign-off on design releases, (3) novel structures used in an ad hoc way for integration, (4) job rotation practiced in engineering functions, and (5) permanent reassignment across functions. New structures that included effective teams for integration accounted for a significant amount of the return on investment on these modernization projects. Like Salzman (1990), Ettlie (1990) found that the minority (15%) of firms involve work force participants in the new system design process. However, Ettlie also found that these firms are the most innovative. A question that this line of work raises is what, if any, optimal combination of people and skills promotes the effectiveness of the concept development phase of development.

Several cases have surfaced recently that illustrate how the early stages of the R&D process can be actively managed to impact later developments. In particular, Naj (1990) presents the General Electric case, which uses "cross-pollination" of laboratory personnel who become intimately involved in strategic issues of the business. Not only do bench scientists become involved in takeover strategies, they are now required to be entrepreneurial.

THE CONTEXT OF CONCEPT DEVELOPMENT

Weiss and Birnbaum (1989) argue that the ability to implement a technological strategy (long-term program of technological change) of an organization depends in great measure on the organization's infrastructure—especially relationships with suppliers. Hayes, Wheelwright, and Clark (1988) include within the infrastructure of an organization such management policies and systems as human resource policies, quality systems, production planning, new product development, reward systems, and organizational structure (p. 21).

New products and new processes are not conceived in a vacuum. They unfold in organizational settings of groups of people organized by functions. Not only would the immediate integration of functions be involved in development processes (Ettlie, 1990) but the integration context of development would be logically implicated as well (Lawrence and Dyer, 1983). Therefore, such coordination involving the information function of the organization, the relationship between professional staff and line operatives and support (e.g., maintenance), and the general use of teams in the unit would have an important contextual impact on concept development. It likely would promote more ideas' being brought to the attention of groups of decision makers because integration is promoted in the unit. The context of concept development is explored in this study and is discussed further under propositions below.

CONCEPT DEVELOPMENT EFFORT AND PROJECT SUCCESS

The remainder of this chapter presents a study that provides some support for the importance of concept development in new product-process development projects. The study shows that the amount of concept development effort is related to the successful introduction of new project and process technologies. The support demonstrated by this study needs to be qualified, however, in that it primarily focuses on the R&D-manufacturing interface rather than the R&D-marketing interface and focuses on projects in which the introduction of new manufacturing processes is as critical or more critical than the introduction of new products. Furthermore, the measures of success are based primarily on manufacturing criteria. Nevertheless, the results of the study are instructive for understanding design-manufacturing integration because new products and processes are often introduced at the same time. Also, manufacturing-based criteria reflect success during the project in matching product and process parameters.

Propositions

Based on findings from the new product development literature suggesting that firms have undervalued early efforts systematically and formally to screen new product ideas, it is suggested that increased emphasis on concept development will promote the success of modernization projects. This proposition can be stated as follows:

Proposition 1: The greater the amount of effort allocated to concept development during planning for modernization projects, the more successful these projects will be.

Based on the emerging design management literature, it appears that the more organizational mechanisms are used to promote the integration of core disciplines, the more successful the new product-process development will be. Further, Van De Ven and Delbecq (1976) found that as the uncertainty of the tasks undertaken by the work group increases, the use of impersonal coordination decreases, while personal coordination (e.g., supervisor) and horizontal communication increases. One of the consequences of increasing integration is that representatives from various disciplines on the design team will have to learn other viewpoints and perspectives. What is more, as these new perspectives are introduced into the design process, more time will be spent on concept development. The proposition can be stated as follows:

Proposition 2: The greater the integration between core disciplines (i.e., design and manufacturing) during the design process, the more effort will be devoted to concept development.

The context of concept development, it was argued, is likely to have a significant influence on the amount of effort that is considered to be legitimate in any organizational setting. What is more, technical context (e.g., information systems) can be an asset in concept development by providing coordinated information on standards, what is feasible for implementation, and what can be supplied from outside the organization or from some other unit within the firm. The more integrated this context (Lawrence and Lorsch, 1967), the more likely concept development will have an efficient backdrop for innovative proceedings. The following proposition is offered.

Proposition 3: An integrated context for concept development will promote increased levels of effort at the early stages of new product generation.

Finally, two issues are explored in this study. The first issue is that perhaps there is an optimal level of effort that needs to be allocated to concept development depending on various circumstances and conditions of the particular project like resources, scope, and urgency. If concept development has been traditionally underemphasized, perhaps just a linear relationship will be found. Second, the issue of the ratio of design to manufacturing engineers and its relationship to concept development effort is explored. One might speculate that organizations dominated by design engineers might spend less time in concept development because fewer manufacturing engineers become seriously involved in design at the concept development stage. These manufacturing issues are likely to be raised later in the development process.

Sample

The data for this study were generated from a cross-sectional panel of firms and interviews with project personnel in modernizing plants during 1987. The population for the study was defined as all domestic plants undergoing significant modernization of manufacturing processes during our sampling time frame (1983–1987). All firms, plants, and individuals were guaranteed anonymity.

The sample was compiled by merging lists of announcements of significant, multiple component, flexible manufacturing and assembly systems from several sources. Trade publications such as *Automation News, Metalworking News,* all the major robotics publications, and *American Machinist* were screened for announcements of system purchases. We expected at least a 50% response rate, so we worked until we had a list of about 80 cases. A target sample of approximately 40 plants was assumed to be adequate based on size effect (detectable influence of causal variables) for this type of study (Cohen, 1977).

In order to be eligible for the study, a firm was required to have committed resources for the purchase of the system, although not necessarily to have installed the system as yet. A plant was the primary unit of analysis of the study. A total of 39 (66%) of the first 69 eligible plants we contacted agreed to participate in the study for the first panel data collection. Non-response bias and independence of plant cases that shared the same firm were evaluated and no significant results were obtained. Further, we found no significant industry effects based on one-way analysis of variance by Standard Industrial Classification.

In the resulting sample of 39 plants, all regions of the country were represented. The systems purchased were typically either flexible manufacturing systems, defined in very broad terms to include multiple-machine, computer-integrated, materials-handling-intensive, discrete-parts-producing systems or flexible assembly systems (19 or 49%). Median new system cost was $3.6 million, but 10 (26%) of the plants spent less than $1 million initially on hardware and software.

Participating plants were operated primarily by large firms, with the vast majority (34 or 87%) having more than 500 employees. Plants also tended to be large, with about 60% having more than 500 employees on site. The smallest plant had 12 people, the largest had 12,000, and the median plant had 850 employees.

Respondents

Over 100 personnel were interviewed in plants for the study. One primary respondent was identified in each case, although we usually interviewed two or three people at each plant. Top managers (20) and middle managers (10) were the primary respondents and accounted for 76.9% of the sample of respondents.

Design-Manufacturing Integration Measure

Extensive preliminary data collection in both the technology supply vendors and firms that have used robots and flexible manufacturing systems, as well as a review of managerial practice and technology adoption trends in the durable goods industries under study (cf. Graham and Rosenthal, 1986) produced candidate questions for the interview schedule. Thirty-five of these questions and items focused on the design-manufacturing integration dimension.

The general approach to developing scales here was to limit questions to behaviors and actions that could be verified on plant tours. Therefore, the response format in interviewing was generally limited to yes (the action had been accomplished), no, or in process. When "in process" was the response, a question was asked to indicate how plans were proceeding to implement the action. These responses were coded 3,

1, and 2, respectively. In all cases, and again for each section on the interview schedule, respondents were reminded that these practices had to be tied specifically to the modernization program and system under study, not just activities in general. In many cases respondents were required to give examples of the practice in order for their response to be qualified as a yes.

The scale to measure design-manufacturing integration resulted in the inclusion of five items (Cronbach alpha = .73) with an average inter-item correlation of r = .36 (n = 27) and a scale mean of 9.15 (s.d. = 2.92). The design-manufacturing items were as follows: (1) "We have people who are trained in DFA (design for assembly) or DFM (design for manufacturing)." "If the answer is yes—Who was trained for what?" (2) "A manufacturing representative is required to sign off on design reviews for new products on this system." (3) "We have developed and implemented new structures in order to coordinate design and manufacturing." "If the answer is yes—What are these structures?" (4) "Job rotation between design and manufacturing engineering is practiced in this firm." (5) "Personnel from design engineering are sometimes moved to manufacturing or vice versa."

The most rare modernization-related behaviors, based on interview reports and examples within this scale, are job rotation between design and manufacturing engineering, with only three (7.7%) of the plants agreeing with that statement, and the use of training in DFA or DFM, with only six (15.4%) of the plants reporting that practice as part of their modernization program. Movement of engineers on a permanent basis between design and manufacturing is apparently a more common practice, although still relatively rare, with 13 (33%) of these plant representatives reporting the practice. Typical examples include two cases of senior manufacturing engineers in automotive plants who were transferred to product teams.

The ratio of design to manufacturing engineers was documented during interviews. The average was 3 to 1, but no normative data are available to see if this is typical. A ratio of 2 to 1 maximizes the integration-scale sum (Ettlie, 1990).

Technology Radicalness

Radicalness of process technology incorporated into each new manufacturing (vs. product) system was determined by a panel of expert judges using the following protocol: (1) Radical technology is a rare event; e.g., typically only 10% of all new products introduced each year are new to the world (Booz, Allen, & Hamilton, 1982). (2) Radical technology incorporation involves new science that is a demonstrable departure from existing practice, e.g., cutting metal bars with lasers versus saws. (3) New technology that is radical is risky to adopt because of lack of precedent in use, and the outcomes in practice are uncertain (Hage, 1980). (4) Radical process technology requires new skills and attitudes to deploy, and a learning curve effect is observed when it goes into use. For the radical category alone, and for a panel of six engineers, there was a 77% agreement with the author on categorization of first panel cases.

In this coding exercise, and before case data were presented to these engineers, the judging group agreed on five additional criteria for the protocol specific to process technology: (1) level of integration of the system; (2) flexibility; (3) machine intelligence; (4) robustness—or degree of effective operation in a degraded performance mode; and (5) the number of new features incorporated into a system. Eight (20.5%) of the 39 cases were coded as radical technology systems by the first panel and five (12.8%) by the third panel after replacement for attrition. A more limited group of the same judges was used to check categorization of the replacement cases with continued high (over 85%) agreement on type of technology incorporated in the systems studied.

Concept Development Effort

The amount of effort expended on concept development was measured by the percentage of total project effort devoted to this stage. The entire section on design-manufacturing integration in the interview schedule was introduced by the following introduction, which was read to all respondents:

> Next, we will be reading a set of statements which describe activities that may be used to coordinate the design and manufacturing functions. As part of your system's deployment strategy, please indicate if you use any of the following activities to coordinate the design and manufacturing functions. You may call these programs simultaneous engineering, concurrent engineering, process-driven design or similar term.

The specific question on concept development was, "What percentage of a project's duration is devoted to concept development? (answer in %)." That is, the respondent was asked to comment on general practice, as a context for the specific new system's deployment being studied and relative to the products that typically flow through production systems at that site. This refers to the amount of effort expended on new and improved products for the specific new manufacturing system in question. It varies widely in these manufacturing units. The range starts at a low of 1% and goes to a high of 66% based on a sample of 25 reporting cases of in-plant survey data. The mean level of effort was about 28%, which was nearly equivalent to the median of 30%, so the distribution does not appear to be skewed. The remainder of the analysis is restricted to these 25 complete data cases.

Integrated Context for Concept Development

In the same interviews, we asked questions about the context of modernization, and a pool of questions concerning teamwork and bridging functions was scale-analyzed as in the case of the design-manufacturing integration measure (using the same format: yes, no, and in process, with scores of 3, 1, and 2, respectively). The following four items resulted: (1) "We have formed meetings and teams to resolve the MIS-CIM (management information system-computer-integrated manufacturing) interface." (2) "Our plant structure tends to be layered, with top down controls." (3) "We have engineer and blue-collar teams." (4) "We have productivity teams." The corrected item-total correlations for this four-item scale ranged narrowly from .46 to .40. The Cronbach alpha was .645, and the standardized item alpha was .647, which is acceptable for just four items. (Cronbach's alpha tends to be inflated by increasing the number of items.) The scale mean was 9.31 with a standard deviation of 2.6, which is slightly higher than the theoretical scale mean of 8 (2 x 4).

The scale could be called the "integrated context" of the concept development process since three of the four items related to integrating functions and levels within the firm. Item 3 (engineer-blue-collar teams) was also correlated with an index of administrative innovation and within the setting of a layered plant hierarchy represents potent horizontal communication capability. The MIS-manufacturing interface has been identified as a critical spot in most modernizing plants, so it is a key item on this scale.

Dependent Variables

A total of seven performance measures was included in the interview process and produced sufficient data to be analyzed here. These included primarily capacity-enhancing performance measures: scrap and rework, percentage of target cycle time achieved, uptime, utilization, cost per part reductions, and system turns. Three of these measures (percentage of target cycle time achieved, utilization, and uptime) were validated using expert opinion. A rating of the overall success of the integrative effort was requested also.

Results

The correlations between the percentage of total project time allocated to concept development and the performance measures included to evaluate the success of these modernization programs appear in Table 7.1. Missing data were deleted pairwise for variables. The first proposition states that the greater the amount of effort allocated to concept development, the more successful modernization projects will be.

Proposition 1 is supported for some but not all of the performance measures included here. Concept development effort is significantly correlated with new system utilization ($r = .35$, $p = .055$), cost per part reduction ($r = .67$, $p = .034$), system turns ($r = .82$, $p = .006$), and the opinion rating of the overall success of the integration effort ($r = .37$, $p = .05$). The number of observations for turns and cost reduction is small, and there is no apparent simple linear relationship between concept development effort and typical production performance measures like cycle time and uptime. The relationship between concept development and scrap & rework (as a percentage of total manufacturing cost) is in the predicted direction but not statistically significant ($r = -.32$, n.s.).

Given these mixed results for proposition 1, the second-order effects of concept development were tested to evaluate simple nonlinear relationships. The results of these analyses appear in Table 7.2. Eta versus Pearson r comparisons were done to select candidates for second-order testing (Eta $\gg$ r).

The tests for second order effects show that the only variable tested that appears to exhibit any curvilinear effects with concept development effort is organization size as measured by the number of year-round firm employees. The first-order regression accounts for about 3% of the variance (R^2) in concept development effort and the second-order regression accounts for about 17%. This amounts to an increased explained variance of 14%, but the difference just fails to be statistically significant on traditional criteria ($F = 3.89$, df $= 24$, $p = .06$). No outliers were deleted. Apparently, medium-sized manufacturing firms (about 5,000 to 22,000 employees in this

Table 7.1. Correlates of Concept Development Effort

	Design manufacturing integration	Integrated context	Ratio of design to manufacturing engineers	Scrap & rework	Cycle time	Uptime	Utilization	Cost per part reduction	System turns	Success of integration effort
Concept development effort[a]	.40[d] (n=17, p=.057)	.61 (n=22, p=.001)	-.49[c] (n=20, p=.014)	-.32 (n=16, p=.114)	-.27 (n=20, p=.122)	-.15 (n=22, p=.25)	.35[d] (n=22, p=.055)	.67[c] (n=8, p=.034)	.82[b] (n=8, p=.006)	.37[c] (n=21, p=.05)

[a]Measured as a percentage of total project duration for new and improved products for the new manufacturing system.
[b]p < .01.
[c]p < .05.
[d]p < .10.

118

Table 7.2. Summary of Second Order Effects for Concept Development Effort (Dependent variable)

Independent variable	R^2		$F(df)$[a]	P[b]
	First order	Second order		
Design-manufacturing integration	0.044	0.162	1.98(df = 16)	p = .24
Number of year-round plant employees	0.027	0.089	2.09(df = 24)	p = .18
Number of year-round firm employees	0.027	0.173	3.89(df = 24)	p = .063
Scrap & rework	0.102	0.147	0.69(df = 15)	n.s.

[a] These values interpolated from Table D, in *Statistics for Experimenters*, George E.P. Box, New York, John Wiley & Sons, 1978.
[b] This is a test of the significance of increased explained variance for the 2nd-order regression.

sample) spend the most time proportionately on concept development for modernization projects that require process and product changes. Yet most would consider these firms "relatively" large by other standards (e.g., the Small Business Administration uses 500 employees as the cutoff for small firms).

Proposition 2 states that the greater the integration between disciplines involved in design, the greater the effort devoted to concept development. The correlation (Table 7.1) between design-manufacturing integration and concept development effort was r = .40 (n = 17, p = .057), which suggests that this proposition is weakly supported. Then the radicalness of the technology incorporated into the new system was controlled and the partial correlation for design-manufacturing integration and concept development effort was r_p = .46 (df = 11, p = .058). Apparently, the radicalness of technology does not diminish the strength of the relationship between these two factors, although the causal direction of the pair needs to be determined.

Proposition 3 states that an integrated context (use of cross-functional coordinating mechanisms) for concept development, although it does not involve design and R&D directly, will promote increased effort in early stages of product-process development. The results for testing the proposition are included in Table 7.1. The correlation between concept development effort and integrated context was r = .61 (p = .001, n = 22), which strongly supports proposition 3.

One might argue that without an integrated context, the luxury of increasing the effort devoted to concept development would not be tolerated (cf. Pugh, 1981). Another explanation for the result is that the same causes of integrated context promote more concept development effort—general management influence would be one candidate. Innovation strategy is another. Integrated context was significantly correlated (r = .34, n = 32, p = .029) with manufacturing technology policy (Ettlie and Bridges, 1987; not shown in Table 7.1), which supports this supposition.

The correlation between the ratio of design to manufacturing engineers (Table 7.1) and concept development effort was r = -.49 (p = .014). This result suggests that as the number of design and manufacturing engineers becomes more balanced in an organization, more effort is expended on concept development for this sample of modernization projects.

DISCUSSION

Concept development has been variously claimed by marketing, R & D, including design; and now by cross-functional teams that claim to be "simultaneously" designing products and processes. The effort expended in concept development appears to lead to some higher-performance outcomes, like reduced costs, and does vary directly with organizational integration between design and manufacturing, and this relationship is not altered substantially when the degree of radicalness of technology incorporated into these new product-process systems is controlled. This linear relationship might have been supported because levels of concept development effort have been traditionally too low in these firms.

The search for "optimal" levels of concept development effort was unsuccessful in this empirical study. Although the typical design team in this sample spent 30% of its time on concept development, this can hardly be prescribed to other situations under these circumstances. Concept development effort appears to depend on other factors. Furthermore, the quality of the stage may be as important as its duration.

The finding that design-manufacturing integration varies directly with concept development effort and that this relationship is still salient when controlling for radicalness of technology used is important. It is typically assumed that when radical new technology in product or process is required, the amount of time in development increases. However, it does not appear to influence the relationship between concept development effort and design-manufacturing integration. Those cases that are high on one are also high on the other. The direction of the causal relationship has not been determined in this research. Other development effort times are typically quite challenging as well. For example, software has chronically been late on these projects, and this factor requires further investigation.

Results indicated that the context of concept development was quite important to the amount of concept development effort expended at early stages of new and improved product design. The more integrated this context (MIS/design/blue-collar/manufacturing), the more effort early on. This suggests, along with other results, that this context is a resource and a supporting mechanism for a tolerance of greater effort at the early stages of new product design during modernization, at least. It is not clear whether this mechanism operates during "normal" periods of the organizational life cycle, if there is such a thing these days in manufacturing.

The finding that "medium"-sized firms appear to spend more effort on concept development is provocative and begs to be refined and replicated. It may be that smaller firms have less trouble integrating functions and larger firms have such great difficulty in coordinating and "reintegrating" teams back into the existing structure that medium-sized firms have the real advantage here. This proposition needs to be tested with a much larger and more diverse sample of organizations. Once one leaves discrete-parts industries, a whole new set of design issues emerge as well.

The role of suppliers was not explored in this study and needs to be included in the new research of this type. Since it appears that the ratio of design to manufacturing engineers is related to concept development (the greater the balance, the more effort on concept development), the role of suppliers should be investigated also. "Pure" new product projects ought to be compared to product-process projects, which were in this sample.

The other avenue to explore is, of course, the impact of customers or the "voice" of the customer in design. This includes the role of marketing and sales in core team evolution. Yet, we know little of how the voice of the customer is actually incorporated into concept development and how this influences time to market with new products. It could act variously. On the one hand, it could speed product introduction by impressing urgency. On the other hand, it could increase response time by adding complexity to decision-making. This topic should be investigated.

ACKNOWLEDGMENT

Work in this area was supported by the National Science Foundation. The opinions in this chapter are those of the author and do not necessarily represent the official opinions of the NSF. The graduate student on this project was Deborah Schut.

REFERENCES

Baker, Norman R., Stephen G. Green, and Alden S. Bean. "How Management Can Influence the Generation of Ideas." *Research Management,* November-December 1985, 35–42.

———, J. Siegman, and A. H. Rubenstein. "The Effects of Perceived Needs and Means of the Generation of Ideas for Industrial R&D Projects." *IEEE Transactions on Engineering Management* EM-14(4), December 1967.

Bar, Jacob. "A Systematic Technique for New Product Idea Generation: The External Brain." *R & D Management* 19(1), 1989, 69–78.

Booz, Allen, and Hamilton, Inc. *New Products Management for the 1980's.* New York: Booz, Allen, and Hamilton, Inc., 1982.

Cohen, J. *Statistical Power Analysis for the Behavioral Sciences.* New York: Academic Press, 1977.

Constandse, W. J. "Why New Product Management Fails." *Business Management,* June 1971, 16–19.

Conway, H. Allen, and Norman W. McGuinness. "Idea Generation in Technology-Based Firms." *Journal of Product Innovation Management* 4, 1986, 276–291.

Cooper, R. G. "Overall Corporate Strategies for New Product Programs." *Industrial Marketing Management* 14, 1986, 179–193.

Crawford, C. Merle. *New Products Management.* Homewood, Ill.: Irwin, 1983.

Eppinger, Steven D., Daniel E. Whitney, Robert P. Smith, and David A. Gebala. "Organizing the Tasks in Complex Design Projects." Working Paper #3083–89–MS, MIT Sloan School of Management, June 1990.

Ettlie, J. E. "Policy Implications of the Innovation Process in the U.S. Food Sector." *Research Policy* 12, 1983, 239–267.

———. "Methods that Work for Integrating Design and Manufacturing." In *Managing the Design-Manufacturing Process,* eds. J. E. Ettlie and H. W. Stoll. New York: McGraw-Hill Book Company, 1990, 53–77.

——— and W. P. Bridges. "Technology Policy and Innovation in Organizations." In *Technology as Organizational Innovation,* ed. J. Pennings and A. Cambridge, Mass: Ballinger Publishing Company, 1987, 117–137.

——— and R. D. O'Keefe. "Innovative Attitudes, Intentions, and Behaviors in Organizations." *Journal of Management Studies* 19(2), April 1982, 153–162.

———, W. P. Bridges, and R. D. O'Keefe. "Organizations Strategy and Structural Differences for Radical versus Incremental Innovation." *Management Science* 30, 1984, 882–95.

Grabowski, H., and J. Vernon. "A New Look at the Returns and Risks to Pharmaceutical R&D." *Management Science* 36(7), July 1990, 804–821.

Graham, M. B. W., and S. R. Rosenthal. "Flexible Manufacturing Systems Require Flexible People." Manufacturing Roundtable Research Department Series, Boston University, 1986.

Hage, J. *Theories of Organization: Form, Process, and Transformation.* New York: Wiley, 1980.

Hayes, Robert H., S. C. Wheelwright, and Kim B. Clark. *Dynamic Manufacturing,* New York: The Free Press, 1988.

Hoglund, William E. "Product Program Management Process at GM." Presented at the First Klein Symposium on the Management of Technology, August 24, 1990, University Park, Penn.: The Pennsylvania State University.

Jelenik, M., and C. B. Schoonhoven. *The Innovation Marathon.* Oxford, England: Basil Blackwell Ltd., 1990.

Lawrence, P. R., and J. W. Lorsch. *Organization and Environment.* Boston: Division of Research, Graduate School of Business Administration, Harvard University, 1967.

Lawrence, P., and D. Dyer. *Renewing American Industry.* New York: The Free Press, 1983.

Levin, Doron P. "Hot Wheels." *The New York Times Magazine,* September 30, 1990, Section 6, 32–33, 72–78.

Lilien, Gary L., and Eunsang Yoon. "Determinants of New Industrial Product Performance: A Strategic Reexamination of the Empirical Literature." *IEEE Transactions on Engineering Management* 36(1), February 1989, 3–10.

Mowery, D., and N. Rosenberg. "The Influence of Market Demand Upon Innovation: A Critical Review of Some Recent Empirical Studies." *Research Policy* 8, 1979, 102–153.

Naj, Amal Kumar. "GE's Latest Invention: A Way to Move Ideas from Lab to Market." *Wall Street Journal,* June 14, 1990, A1, A9.

Nord, Walter R., and Sharon Tucker. *Implementing Routine and Radical Innovations.* Lexington, Mass: D.C. Heath and Co., 1987.

Ottosson, Stig. "Guided Product Idea Generation." *OMEGA, The International Journal of Management Science* 11(6), 1983, 547–557.

Pinto, Jeffrey K., and Dennis P. Slevin. "Critical Success Factors in R & D Projects." *Research & Technology Management,* January-February 1989, 31–35.

Pugh, Stuart. "Design Decision: How to Succeed and Know Why." *Small Piece Reader in Engineering Design.* Loughborough, England: University of Technology, DES 1981.

Ram, Sundaresan, and Sudha Ram. "Expert Systems: An Emerging Technology for Selecting New Product Winners." *Journal of Product Innovation Management* 6, 1989, 89–98.

Rothwell, R., and T. G. Whiston. "Design, Innovation and Corporate Integration." Presented at the International Conference on R&D Design and Manufacturing, University of Ghent, Belgium, September 12–13, 1989.

Salzman, Harold. "Designing Process Technology for Usability: Practices and Principles for Strategic Design of Skill-based Technology." Presented at Technology and the Future of Work, Stanford University, March 28–29, 1990.

Schoonhoven, Claudia B., Kathleen M. Eisenhardt, and Katherine Lyman. "Speeding Products to Market: Waiting Time to First Product Introduction in New Firms." *Administrative Science Quarterly* 35, 1990, 177–207.

Smith, D. "Sequential Development, Simultaneous Engineering, and Simultaneous Development: Distinctions, Methods and Benefits." Draft Working Paper, Industrial Development Division, Ann Arbor, Mich.: University of Michigan, 1990.

Somers, Mark John, and Dee Birnbaum. "Managing Product and Production Technology: A Two Stage Strategic Challenge for the High Technology Firm." *Proceedings, Strategic Leadership in High Technology Organizations,* Boulder, Co., January 1990.

Souder, William E. *Managing New Product Innovations.* Lexington, Mass: D.C. Heath and Company, 1987.

———. "Improving Productivity Through Technology Push." *Research/Technology Management* 32(2), March/April 1989.

Trygg, Lars. "The Use of Integration Mechanisms in the Design to Production Transfer." Working Paper No. 1990:31, presented at the International Conference on Advances in Production Management Systems, August 20–22, 1990, Espoo, Finland.

Urban, Glen L., and John R. Hauser. *Essentials of New Product Management.* Englewood Cliffs, New Jersey: Prentice Hall, 1987.

Van De Ven, Andrew H., and Andre L. Delbecq. "Determinants of Coordination Modes within Organizations." *American Sociological Review* 41, April 1976, 322–333.

Weiss, Andrew R., and Philip H. Birnbaum. "Technology Infrastructure and the Implementation of Technological Strategies." *Management Science* 35(9), August 1989, 1014–1026.

Welsch, M. Ann, Gordon E. Dehler, and Stephen G. Green. "A Strategic View of Project Initiation." Presented at the Academy of Management Meeting, San Francisco California, August 12–15, 1990.

Wind, Yoram, and Vijay Mahajan. "New Product Development Process: A Perspective for Reexamination." *Journal of Product Innovation Management* 5(4), December 1988, 304.

Zinger, Billie Jo, and Modesto A. Maidique. "A Model of New Product Development: An Empirical Test." *Management Science* 30(7), July 1990, 867–883.

8

PROTOTYPES FOR MANAGING ENGINEERING DESIGN PROCESSES

E. ALLEN SLUSHER and RONALD J. EBERT

Product engineering design (PED) is widely recognized as an important component in the competitive responses of manufacturing firms. Accordingly, major advances have occurred in many technical areas of design, for example, CAD, expert systems, embodiment design, and Taguchi methods. However, the benefits inherent in these tools are often unrealized because the overall management of the design process is not tailored to the special demands of each project. There is a tendency for project managers to use the same basic management approach for all projects. Although good managers intuitively make adjustments for the special demands of each new project, they must rely on their own limited personal experience. The purpose of this chapter is to improve engineering design management by developing a conceptual framework for classifying design situations and to show how the framework can be used to select a management process appropriate for a given project. Although each project must be uniquely managed, it is possible to identify four prototypical design situations and to make general recommendations for managing each type.

In our framework the design process is viewed as a dynamic activity involving the interaction of technical, behavioral, and organizational factors. Overall success in design projects depends heavily on appropriate information flows—sources, timing, intensity, transmission modes, quality, and quantity—among diverse specialists and managers. The nature of these information flows, their effect within PED, and the ways they should be managed are contingent on the requirements of the design task and the requisite coordination with other organizational units (e.g., R&D, marketing, and manufacturing) and external stakeholders (e.g., suppliers, contractors, regulators, consultants, and customers).

Despite the dramatic impact that engineering design has made in creating competitive advantage, American industry has devoted less effort to understanding the PED process than Japan or Germany. PED practices in America have not been extensively studied or well documented. In commenting on their design management choices, practitioners usually cite historical precedent (we've always done it this way), experience (it seems to work), or trial-and-error (we try it one way for a while, then we try another). American manufacturers often reinvent PED practices used in other firms or manage projects using ad hoc procedures. The absence of any unifying conceptual frameworks for PED has resulted in widely different approaches to managing the design process and consequent variances in PED effectiveness. The focus has been on specialized engineering sciences with less attention to articulating an integrated product development process in which PED plays a crucial role. Large firms often implement structured approaches to the design process that identify major design phases, control procedures, and planned iterations. However, these structured approaches usually do not recognize the qualitatively different design situations that arise. In practice, structured approaches are applied primarily in product-variant redesigns. When faced with PED projects involving a major "clean-sheet" design, structured approaches are often misleading and inadequate. The present framework gives guidance for creating more appropriate management approaches for each prototypical design situation.

Design projects vary in their relative importance, expenditure levels, and management structure. Moreover, these differences and the methods for addressing them are seldom codified. There is usually no systematic documentation on the relationship of design management processes to project effectiveness. In addressing the perplexing difficulties in managing the design process, we have found that an information-processing framework provides a useful conceptual perspective on PED. Many organizational factors act as either facilitators or barriers to the flow of information in PED. The organization structure (including departmental boundaries), reporting procedures (formal and informal), centralization of decisions, technical and functional specialization (with specialized languages and encoding), the design project's organization (functional segmentation), and project uncertainty (scope, visibility, and goals) all either deter or promote the information flows that ultimately determine project success.

Information processing serves the dual purpose of reducing equivocality and providing a sufficient amount of information for performing the tasks required in PED (Daft and Macintosh, 1981; Daft and Lengel, 1986). Information amount can be measured by the volume of data that is gathered and interpreted for various design activities. In a project setting, it involves seeking, acquiring, utilizing, synthesizing, and transferring information. Information is transferred between design specialists to provide technical and organizational information for their tasks. Information is also brought into the project from the larger organization and from sources in the external environment including consultants, competitors, and customers.

Information equivocality, another vital concept in our framework, refers to the multiplicity of meaning or ambiguity conveyed by information about design activities (Weick, 1979). Unequivocal information has a clear and uniform interpretation among its users, whereas equivocal information has multiple or conflicting interpretations. Equivocal information can impair project effectiveness because the sequential nature of PED imposes interdependencies—performance at one stage depends on information from previous, concurrent, or successor stages—in the design project. Ambiguous information, especially from qualitative and judgmental upstream stages, fosters downstream delays. Inadequate needs identification, for example, leads to confusion and delays in prototype development. Several techniques have been developed for reducing information equivocality, including quality planning charts for clarifying needs requirements (Sato, 1983) and Pugh's (1981) procedure for the early translation of needs into concepts that are conceptually invulnerable. Similarly, the clarity

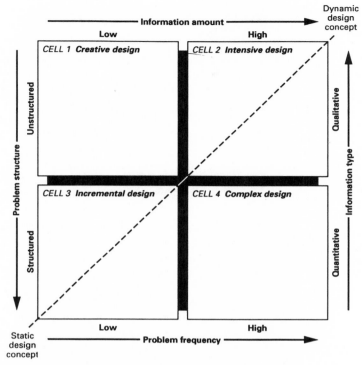

Figure 8.1. Engineering design prototypes.

of the design objectives for such downstream stages as process engineering is critical information. When design objectives are equivocal, duplication of effort and time delays are created. The downstream technical expertise in specialized departments is wasted instead of being beneficial. With ambiguous or conflicting objectives, design engineers will misunderstand how their specialty tasks relate to other project activities. Engineers may lock into "one best way" of practicing their specialized skills when, instead, the best way depends on objectives that are unique to each project. For example, structural analysis of one component may require a highly detailed and refined analysis while an alternative component may require only a "rough-cut" level of detail for the project's current stage of development. The ability to tailor information requirements differently for various design tasks is facilitated by an understanding of how overall information requirements vary within prototypically different types of design projects.

A CONTINGENCY MODEL OF PED

Although design managers and engineers intuitively recognize that information needs will vary for different design projects, they have no articulated framework for guiding their decisions regarding information management. In response to this void, we have developed a typology for classifying design projects (see Figure 8.1). The typology identifies two pivotal factors, the structure of the design problem and the frequency of novel problems facing the designers (see the problem structure and problem frequency dimensions in Figure 8.1). Our framework is distilled from our observations of PED practice and from research conceptualizations in the literature on business management. Prescriptive statements about how design projects should be planned, organized, and led in four prototypical situations are then derived from the model.

In our typology, the design process is visualized as an information management activity. Managing design is a process of managing information flows through various stages of the design process. The crucial questions are, How much information do we need? When do we need information? What kind of information do we need? How do we obtain information? How do we transmit information? The answers to these questions depend on the specific design task assigned to the project team.

We have found it useful to recognize two reservoirs of information that project teams utilize in any design environment. The first source is the team's personal knowledge gained from its own experiences. Another source is general design knowledge contained in books, computer databases, organizational documents, and in the minds of designers. Faced with any new assignment, project managers and team members are likely to first rely on their own previous design experiences. When the current problem is recognized as similar to a previous successfully solved problem, a team's first impulse will be to employ the analytical framework used in the previous project. Even this seemingly simple information retrieval task has barriers. Because designers may not recognize similarities between past problems and the current problem, a transference opportunity may be lost. If the current problem lies outside the team's existing repertoire, they will extend the search and explore other standard information sources. Only when familiar sources have been exhausted will the team expend additional effort to identify and exploit new information sources. The barriers to finding relevant information become greater as the search moves from familiar to unfamiliar information sources. Knowledge constraints, time constraints, cost constraints, and access requirements act to restrict the information search and the sources that will be tapped. In practice, the design team's search for information is constrained to the accumulation of satisfactory amounts rather than maximum amounts of information (Nevill and Paul, 1987).

A Typology of Design Situations

Incremental design. What determines the amount and type of information a design team needs to complete a project assignment? Clearly, the answer depends on the nature of the design task (i.e., problem structure) and the amount of novel problem solving (i.e., problem frequency) it imposes on the designers. Consider, for example, the design prototype identified as "Incremental design" in cell 3 of Figure 8.1. Here, the design task is to modify existing components or products, retaining the established static design concept. We describe these well-defined tasks as structured because the design variables and their relationships have been established. Since the new design task is similar to past designs, it is likely that considerable, potentially useful, information has accumulated. To preserve it, much of the design knowledge may be systematically retained in written procedures, specifications, drawings, manuals, databases, models, and computer software. This type of design knowledge is precise, usually quantitative, and readily accessible.

When designers can access previous designs that have been precisely documented, there will be a relatively low need for collecting new information. For these structured problems, there is more reliance on quantitative than on qualitative information. The firm can convert existing information into retrievable form that is accessible readily to the design teams (Middendorf and Wang, 1987). Most problems that occur involve only one or a few components or subsystems. Incremental design processes are routine, rather than novel, and relatively low levels of information processing are needed. Problem-solving utilizes the existing quantitative and visual information that is present in organizational databases. Problem frequency is low because of the relative simplicity and small scope of the design tasks. As later design stages are confronted some problems may arise, but they seldom involve novel issues. In summary, incremental design situations are characterized by a known product or process structure and low problem frequency when executing the design task. Therefore, they usually require relatively low amounts of information, primarily quantitative, for their successful completion.

In identifying the remaining three prototypical design situations, the model (see Figure 8.1) treats problem structure and problem frequency as the independent variables. In doing so, it is important to remember that we are examining "relative" differences in information amount and type. Clearly engineering design, when compared to most organizational tasks, is information intensive. However, some design situations will require more information than others. Furthermore, both types of information—quantitative and qualitative—are required for all design situations. Although quantitative information offers the advantages of precision, it often fails to convey enough information to complete the design task. Qualitative information, particularly face-to-face meetings or telephone conversations, provides a "richer" form of communication necessary for solving design issues impacting resource commitments, motivation, and political issues (Daft and Lengel, 1984). The contingency management issues that must be resolved are the amount of information needed and the relative mix between quantitative and qualitative information necessary for each design task.

Complex design. As the design task changes from one part, to subassemblies, to major systems, to total products, we view the design task as increasing in complexity. Complex designs usually require the participation of more people. These additional people greatly increase coordination requirements (Tushman, 1978). Even when each of the component tasks in the overall design project is highly structured, there remains the need to coordinate the design process so that all subtasks will be integrated into a coherent final product design. Increased complexity in the total design process results in more frequent problems, particularly problems resulting from the integration of component designs. There will be uncertainty in managing the design process, but because the design of each component is highly structured the uncertainty is moderated by the large amounts of information that can be transmitted in standard (often quantitative) forms. This "Complex design" situation is represented by cell 4 in Figure 8.1.

Creative design. If the design task focuses on a new component or a relatively simple new product, the designers cannot rely on existing information derived from previous projects. Facing a novel situation, designers must rely on richer, qualitative forms of information to solve their unstructured PED problem. Particularly in the concept development stage, designers are faced with considerable ambiguity in determining the new product concept and its design and development requirements. These requirements cannot be determined solely from quantified data but need qualitative information that includes opinions, judgments, experiences, and commitments from different people and groups within or outside the company. The communication of rich information requires interpersonal contacts between the interested parties. As shown in cell 1 of Figure 8.1, this situation has been labeled "Creative design." The design problem is unstructured and there are no standard procedures that can be applied. By definition, creative designs face high uncertainty in the concept development stage, but the limited scope (e.g., a component or simple product) of the project leads to only a few problems in the later design stages. Since there are relatively few novel problems that arise after design requirements have been specified, the overall information requirements are relatively low.

Intensive design. In cell 2 of Figure 8.1, we have described the situation as one with "Intensive design" processes. Here we have the elements of both the creative and complex designs. There is not only the need to deal creatively with ambiguity regarding new and changing design requirements, but also to cope with more complex products involving many integrated components and subsystems. Intensive design processes have vital interdependencies among several designers, customers, and departments. The effectiveness of the design effort is greatly enhanced by cooperation and integration that facilitate qualitative information flows. One approach is the grouping of specialists together into teams. Alternatively, PED specialists are given work experience in other functional areas. Currently, firms are giving more attention to this issue. For example, at Armstrong World Industries, new engineers work three years in a plant before assuming their engineering assignments (Varljen, 1990). Xerox has used 40% of its R&D personnel on project teams that include engineering personnel (Campbell, 1990). These endeavors are vital for intensive designs because they facilitate open information flows and information processing. Large amounts of information are required to coordinate the many personnel and design activities. Intensive design projects need a high proportion of rich, qualitative information to cope with the uncertainty inherent in the unstructured problem situation.

Hybrid designs. Actual design projects may not fit neatly into a single cell of the prototypes model depicted in Figure 8.1. We have observed some positive correlation between problem structure and problem frequency. As design problems become more unstructured, a larger number of problems are encountered during the design process, and the sheer number of problems leads to a less visible problem structure. Designers are working with a more uncertain and ambiguous product concept. Moving from left to right along the dashed line in Figure 8.1 represents an increasingly dynamic product concept. As a project becomes more dynamic, managers and designers must cope with less structured design assignments involving more frequent problems. If our observations of positive correlations between problem structure and problem fre-

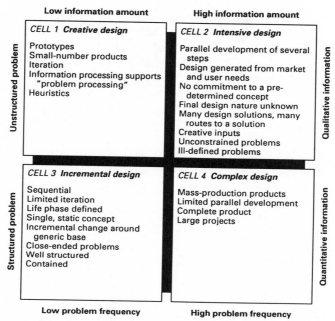

Figure 8.2. Descriptive terms used by design scholars for each prototype.

quency are empirically true, then cell 1 and cell 4 cases will be encountered less frequently in practice than cell 2 and cell 3 cases.

Support for a Prototypes Model

Andreasen (1987) has commented that "we badly need a . . . typology for design situations." Initial support for the usefulness of our prototype model (see Figure 8.1) was obtained by reviewing each of the 124 papers presented at the 1987 International Conference on Engineering Design in Boston, Massachusetts. We found the writings of these design scholars to be supportive of the validity of the independent variables and their relationships in our typology. In Figure 8.2, our interpretations of these scholars' comments on the differences between design situations are classified in the context of the prototypes model. Beginning with cells 2 and 3, which are the "pure" types or true opposites, it can be seen that many labels have been used to describe the design process within each cell. Beitz (1987) describes incremental design (cell 3) as being sequential with limited iterations and defined by the life-cycle phase. Morley and Pugh's (1987) discussion of a single, static design concept with incremental changes around a generic base corresponds to our incremental design situation. Gill (1987) describes this situation as involving closed-ended problems, while Nevill and Paul (1987) use the terms *well structured* and *constrained*. Intensive design (cell 2) is stimulated by market and user needs with no commitments to a predetermined concept; the nature of the final design is unknown (Morley and Pugh, 1987). This situation has open-ended problems (Gill, 1987), unconstrained problems (Nevill and Paul, 1987), and ill-defined problems (Wallace and Hales, 1987). There are creative inputs and many design solutions with multiple routes to each solution (Gill, 1987), and there will usually be parallel development for several steps (Beitz, 1987). The two "mixed" cells (cells 1 and 4) combine characteristics of the two pure cells. Creative design (cell 1) is marked by information processing directed toward "problem processing" (Beitz, 1987). Creative design methods frequently involve prototypes, small-number products, heuristics, and iteration. Complex design (cell 4) focuses on mass-production products requiring some parallel development (Beitz, 1987). These are usually large projects leading to the development of complete products.

SELECTING AN APPROPRIATE DESIGN MANAGEMENT PROCESS

Simply identifying the different design prototypes and their characteristics is insufficient. How should each situation be managed? The emerging interest in this question reflects the growing awareness that PED is too critical for design managers to get locked into "one best way." Instead, there is more willingness to accept the risks inherent in improving their design management processes. Moreover, the risks are lessened because the recommendations discussed below have been tried with success in PED and in the related area of research and development. We offer prescriptive statements for contingently managing design prototypes, even though such statements must be general in nature.

Acceptance of the Contingency Approach

The first step is acceptance of the basic precept that PED projects differ and, consequently, deserve differing types of managerial support. There is growing agreement that design projects should be managed in ways contingent on the special characteristics of the particular project. Andreasen (1987) comments that "Any development project contains a few special features, properties, parameters or conditions that are central for the success of the result, and these elements differ from project to project." Morley and Pugh (1987) note that "The implication is that different types of design activity may require different organizational structures. . . ." Despite the nascent contingency thinking expressed in the engineering design literature, the impression remains that most techniques aimed at improving design are directed toward one design situation, incremental design (Gill, 1987). It may be true that the largest number of design projects are predominantly incremental; however, the importance of these projects to the firm may not be in proportion to their number.

Qualitative Versus Quantitative Information

As design project assignments move from static design concepts to dynamic design concepts (represented by the diagonal line in Figure 8.3), we would expect that larger amounts of information will be required. We also expect that the proportion of richer, qualitative information will increase. The absolute amount of quantitative information may also increase, but at a slower rate. Increases in quantitative information can be handled by sophisticated CAD/CAM systems that integrate three-dimensional design,

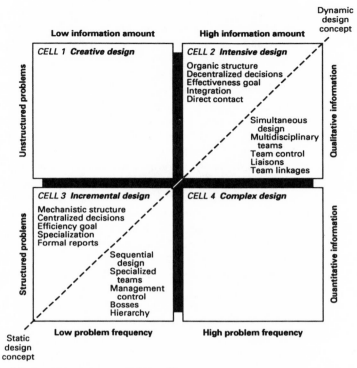

Figure 8.3. Management process contingencies.

two-dimensional drafting, engineering databases, numerically controlled systems, finite-element analysis, and word processing (Primrose, 1987). However, the most crucial current issues in managing the design process focus on the methods used to increase the amounts of qualitative information needed to integrate diverse specialists, departments, and stakeholders.

Mechanistic Versus Organic Management

For incremental design processes there is a static design concept with well-structured problems, and only a few new problems are encountered (see cell 3 in Figure 8.3). In this situation, traditional sequential techniques for engineering design management will be effective. The emphasis is on efficiency and specialization. Design decisions are centralized hierarchically, and top management retains tight control through budgets and formal reports. Design requirements are clearly specified, and engineering is buffered from interference by outside groups. Decisions are made by bosses and not by designers. This management approach is often described as mechanistic and is very efficient for making incremental product modifications (Burns and Stalker, 1961; Ebert, Slusher, and Ragsdell, 1986).

An example of incremental design was experienced by a manufacturer of kitchen appliances. The project was to modify the door assembly of an oven to gain materials savings by reducing the length of the electrical wiring in the door. Another goal was to simplify the door assembly process by modifying the retainer clips that hold the wiring inside the door. The project was assigned to a senior design engineer who, in turn, assigned the wiring and clip subprojects to specialist junior engineers, one a wiring and the other a fastener specialist. The basic concepts of the clip and the wiring system were fixed, as were all related product systems. The junior engineers had access to abundant specialized engineering data on their components and thorough documentation of past designs in their separate work areas.

The engineers faced no significant novel problems in executing their design tasks. The clips and wires were purchased items, and the engineers, working independently, conferred briefly (via telephone and written memos) with supplier representatives for preliminary reactions to engineers' initial redesigns and for cost estimates. The senior engineer imposed deadlines for receiving test versions and deadlines for written proposals for the redesigned components. Once received, the components were test-assembled in the engineering test lab; this was the first time that all three engineers worked face-to-face as a group on the project to ensure their subtasks interfaced properly. After the redesigns were frozen, a manufacturing engineer was brought in to determine the new methods to be used for installing the redesigned components on the assembly line. The modifications were fully implemented in production within six months after launching the project.

Although this was a relatively routine design project, the PED process was efficient and effective. Its result was enduring cost savings in materials and assembly, fewer door problems during assembly, and relatively little disruption from the changeover in manufacturing. The successful design for manufacturability, in this case, arose from the senior engineer's experience base with manufacturing and his comfort with this low-scope, highly structured design task. He centralized the overall control without unnecessary exterior coordination or integration. The quantity and quality of information appropriate for design for manufacturability was thus accomplished by mechanistic administrative procedures. For this incremental design situation, the mechanistic management approach (see cell 3 of Figure 8.3) was both efficient and effective.

Tushman and Nadler (1978) identify two dimensions of project structure that affect a project's information processing capacity. First is its organic-mechanistic nature, which includes the amount of formalization, centralization, and the way in which power, control, and decision-making are distributed. Second are the types of coordination and control mechanisms, including the reward and planning processes, it adopts.

For intensive design processes a more organic management structure is necessary (Burns and Stalker, 1961; Ebert et al., 1986). Initial design concepts are more uncertain, and new problems arise throughout the design process (see cell 2 in Figure 8.3). The emphasis is on effectiveness rather than efficiency and on integration rather than specialization (Naveiro, 1987). Formal reports (paperwork) are replaced by direct interaction (telephone calls, face-to-face meetings). Operational decisions are decentralized, and design teams are given more control over their activities. Multidisciplinary teams are formed within and beyond engineering and perform their tasks simultaneously (Ebert et al., 1986). These teams often include representatives from all areas of the business, especially marketing and production. Where proprietary concerns permit, customers and suppliers may participate in design team meetings.

Organic structures, using greater peer involvement in decisions, less formalization, and less rigid structures are more appropriate for dynamic product/process designs with high interdependence among tasks (cell 2 in Figure 8.3). Consequently, psychological commitment to design decisions is enhanced when control or influence over decisions is more evenly distributed vertically in the organization across engineers, project leaders, and managers (Tannenbaum, 1968). Horizontally (across technical stages) the organic structure encourages joint problem solving and the development of globally shared objectives, rather than locally defined objectives in segmented specialty areas. The organic configuration increases the information-processing capacity on the project by increasing opportunities for feedback, error correction, and synthesis of different viewpoints. In stable project environments with lower task interdependence (cell 3 in Figure 8.3), the organic structure is unnecessary and a mechanistic structure offers more efficient information processing.

An intensive design example occurred when a manufacturer of vehicle engines undertook a major redesign for one of its products. The new product would place the company in a new market that required a radical departure from the existing product concept. A design team's project goals were to create a user-oriented, manufacturable initial design (complete with a prototype) within four months (instead of the standard 24 months). The project involved high levels of technical complexity and inherently interdependent subsystems and components. The team encountered dense information barriers from the specialized languages and grammars that had developed during past years. They had to create a common ground (create a shared language) so they could learn what everyone was trying to do and what they were capable of doing. Everyone participated in defining and sharing specific tasks. A cost analysis specialist dug into finite element analysis on the engineering computer system. A heat transfer scientist worked with a customer team member to understand better the marketing side of the project and its customer needs.

Collaboration among the nearly 30 team members emerged into a "do whatever needs to be done" approach that focused on mutually agreed-upon criteria and objectives, while disregarding any specialist boundaries. Through a process of mutual adaptation during intensive interpersonal interactions, the team members discovered and utilized their individual and group capabilities in a unified manner. As a group they became problem-solvers and decision-makers.

This was a major cultural revolution created by the team and endorsed by top management. Formerly, they were specialists who made recommendations to managers. They were unaccustomed to taking the risks and shouldering the responsibilities for making tough decisions. As a group they agreed to share in the blame and rewards for the outcomes from their decisions. They created a mutually supportive climate of trust that had previously not existed and that facilitated information processing and decision making.

Figure 8.3 contrasts the traditional mechanistic management approach with the newer, more organic, approach. The shifting to more intensive, group-oriented management techniques requires a greater initial expense. For example, Wallace and Hales (1987) found that almost 40% of designers' time was spent in communication in an intensive design project. However, when they are appropriate, these initial expenditures may enable the firm to exploit market opportunities better and prevent costly design errors. Pugh's (1987) Total Design philosophy, for example, is an organic process for insuring that all participants understand how their partial contributions can be integrated into the total design. Applying this philosophy by instituting appropriate design management techniques can make a difference in design effectiveness. Ehrlenspiel's (1987) assertion that "product characteristics can be most easily influenced at the product's inception" is further confirmation of the power in an organic approach in early stages of the PED process. This occurs because designers gain information by obtaining feedback early and frequently from all parties that have a stake in design.

Ehrlenspiel also recommends that an "adviser" from production be appointed to serve as a liaison between production and design. The advisor should be a "highly qualified colleague because he must be acquainted with the demands of design, production, purchasing, and management." An advisor might typically spend one day a week in the liaison role while remaining in the production department. More intense linkage could be created by moving the advisor into the design area. Various teams can be linked by each team designating a member to represent it on another team.

We believe that almost all organizations can benefit from a close examination of their design processes and how they are managed. If one uniform management approach, either the mechanistic or the organic, is taken to all design projects, it is highly unlikely that design will make its full potential contribution to the firm. Similarly, overreliance on one type of information, qualitative or quantitative, for all design projects will deter the product development process. Design management should be contingent on the design process it seeks to control, and the design process itself should be responsive to the competitive product market.

Effectively managing creative or complex design situations requires a flexible management approach. For creative designs (cell 1 in Figure 8.3), organic approaches are needed early in the project to encourage the creativity necessary in developing a product concept. In later stages of design, however, mechanistic techniques will improve project efficiency. In contrast to creative design management, complex designs will benefit from more structured and mechanistic techniques for refining a static design concept. As more organizational groups become involved in later project stages, organic management approaches will be more effective in coping with the many groups involved in implementing a complex design project.

CONTINGENCY DESIGN MANAGEMENT IN PRACTICE

The first step in applying the contingency perspective to engineering design management is deciding which basic design prototype (incremental, complex, creative, or intensive) best represents the current situation. This judgment leads to broad decisions about the overall management structure appropriate for the current project. Having identified prototypical design situations, each requiring a unique management approach, design managers can extend their use of the contingency approach by exploring specific operational issues that may arise in any one of the four design types. In planning their projects, managers should consider the diversity of technical specialties, variability in resource loading, logistical requirements, and multiple stakeholders that will impact the design process. Effective design management not only requires engineering expertise, but also the ability to integrate engineering design efforts with other functional areas, most notably process design, manufacturing, and marketing. Total performance, from initial concept to final design, is determined by formal and informal interactions among various individuals, groups, and departments. How their efforts become integrated throughout the design process depends on the company's organizational characteristics and its business environment (Lawrence and Lorsch, 1967).

Many organizational context factors are significant because they act as facilitators or barriers to the flow of information in the PED process. The organization structure (departmental boundaries), reporting relationships (formalization and red tape), centralization of decisions, technical and functional specialization (specialized encoding and languages), the design project's organization (functional segmentation, integration, team building), and project uncertainty (scope, visibility, objectives) may all deter or promote project success. As we discuss below, PED management should be sensitive to some potential information barriers that are hidden in their project management decisions.

Overcoming Barriers to Effective Information Processing

Two major barriers to information flows are (1) interdependence between tasks and (2) differentiation (or various frames of reference) between project participants (Lawrence and Lorsch, 1967; Slusher and Roering, 1978). As posited by Tushman (1978) and by Daft and Lengel (1984), variations in these barriers will create different information needs. Accordingly, a project's information-processing capacity must be continually adjusted to meet information requirements during the execution of the design. Several methods are available to vary information flows and the challenge is to choose the appropriate method.

Interdependence. We have previously identified the uncertainty arising from a project task that is unstructured and the uncertainty created by frequent problems in executing the design task. Another source of uncertainty is interdependence between tasks within the project. Higher interdependence (as in cell 2 compared to cell 3 in Figure 8.3) requires more effort in coordination and joint problem solving. As required interdependence intensifies, techniques for integration should shift from rules (low-rich media) to mutual adjustment (high-rich media) (Thompson, 1967).

Differentiation. Professional specialists bring to the design project different technical orientations, professional backgrounds, values, goals, and frames of reference (Galbraith, 1977; Lawrence and Lorsch, 1967). They typically use different technical languages and encoding schemes (Katz and Tushman, 1979). As noted by Daft and Lengel (1984), greater coordination among diverse specialists (as in cell 2 vs. cell 3 in Figure 8.3) requires equivocality reduction to overcome different frames of reference and to synthesize dissimilar perspectives into a common view. Vertically in the organization structure, middle managers, project leaders, and technical specialists should employ an appropriate richness of media to create shared interpretations of project objectives and task requirements. Horizontally, the amount of task interdependence among various functional specialists at different project stages will determine the needed amount of information and the necessary emphasis on equivocality reduction.

Problems in coping with differentiation and integration are abundant in engineering design projects. For example, the results of consumer needs analyses are largely qualitative, but they are often conveyed in written documents to specialists in concept selection. The written words in a report often fail to capture the subtlety of the consumers' needs and their implied priorities; therefore, intended adjustments may not materialize at the concept selection stage. Written documentation coupled with verbal clarification (richer media) can lessen this equivocality. In intensive design situations, where differentiation is high, it is usually best to bring the concept selection personnel into the needs-determination stage. Here, the richness of information from face-to-face, first-hand involvement in needs analysis provides an understanding that cannot be conveyed in writing or through conversations among "insiders" to the design process. Concept selection involves intensive, interpersonal information processing (rich media) to clarify selection criteria and to evaluate alternative design concepts and, in this process, creates a common "grammar" among diverse specialists.

CAD/CAM is an extreme example of a technological approach to enhancing information flows and reducing uncertainties arising from task interdependence. Its development is evidence that the high interdependence between engineering design and manufacturing, in some industrial settings, warrants the creation of technologies that enable large amounts of information to be processed and transmitted. Creating CAD/CAM systems involves intensive equivocality reduction through rich media (meetings and phone calls), but once developed, the common grammar that emerges allows the use of less rich media (quantitative and graphical data) to satisfy task information needs. However, CAD/CAM's orientation toward low-rich media may be inappropriate or severely constraining when products and processes are unstable. Dynamic product and process technologies present substantial risks for the decision to rely too heavily on CAD/CAM technologies.

Personnel considerations. In addition to project structure, the orientations of project personnel are also major considerations in information processing. Assigning engineering specialists to projects should be conditioned by informational considerations as well as technical capabilities. Some engineers fit well into specialized activities in a narrow aspect of design with precisely defined task responsibilities and clear statements on the results expected. They may be unconcerned with, or even disrupted by, information on the antecedents for their task responsibilities and how their work blends with subsequent design activities. Such individuals may be most productive in rule-based design environments (cell 3 in Figure 8.3), shielded from what they perceive as extraneous information. Other engineers with similar technical competencies will seek to add a broader project perspective to their specialist responsibilities. They will want to know the relationships between their tasks and the tasks performed by others (e.g., cell 2 in Figure 8.3). If shut off formally from this information, they are likely to initiate search activities to obtain the information informally. One study found that such information was not only sought, but that the choice of information media was related to the uncertainty of the task. For "high-importance" technical information, rich media were selected when uncertainty was high. In contrast, low uncertainty was associated with nonrich media (Holland, Stead, and Leibrock, 1976).

Social motivations. It is important to acknowledge that social motivations influence information flows. Interpersonal dialogue for information processing involves psychological rather than merely technical factors. It provides an avenue for socio-emotional and motivational enhancements (Holland, Stead, and Leibrock, 1976). Formalized information processing, often involving computer systems, enables project members to process and transfer large amounts of technical information with high reliability and validity. However, formal systems do not provide the social support found in interpersonal communication. This is especially important in tasks with high technical uncertainty that carry professional risk for the project member. In these situations, "information-seeking" behavior may really be "social-support-seeking" behavior. The prospects of economic and social sanctions will reduce risk taking and impede information processing. In contrast, selectively involving technical specialists in design activities at stages earlier and later than their own enhances both motivation and information transfer. Thus, we would expect fewer false starts and subsequent task iterations, and a level of design refinement more closely tailored to the requirements of each particular project.

A major challenge in the design process is to mesh human resources with design assignments on projects with different objectives. For example, the initial project objective for designing a new product may be the creation of a first-cut, clean-sheet design sufficient for enabling process engineers to evaluate preliminary manufacturing process requirements against existing manufacturing capabilities and to estimate manufacturing costs of the product. The designers in this situation (cell 2 in Figure 8.3) must (1) be comfortable working with intentionally ambiguous input information; (2) develop an operational understanding of the information requirements of the manufacturing process engineers; and (3) have the personal tolerance for creating a sufficient, albeit incomplete, product design. The information environment in this instance is in stark contrast to a project assignment whose objective is a detailed, CAD-assisted optimization of a product variant (cell 3 in Figure 8.3).

SUMMARY

This chapter has argued that product engineering design should be viewed as a dynamic information-processing activity involving technical, behavioral, and organizational elements interacting throughout the entire design process. Overall project success depends heavily on patterns of information flows among diverse specialists and managers. The nature and management of these information flows is contingent on the design task and requisite coordination among organizational units and stakeholders. A typology for classifying four prototypical design situations (incremental, complex, creative, and intensive) was introduced.

The overall theme of this book is design for manufacturability. Improved DFM requires cooperation between design and manufacturing, aimed at meeting targeted standards for both product and process. Management must decide at what stage in the development cycle to encourage design and manufacturing personnel to negotiate requirements that meet both groups' needs. If negotiations are conducted too soon, changes in the product concept will be uncertain and manufacturing will be unable to plan. If negotiations are conducted too late, the product specifications will be frozen and manufacturing may have to make costly adaptations. The contingency model (see Figure 8.3) presented in this chapter provides some help in timing and structuring interactions between design and manufacturing personnel. The key success factor is determining when and how much the design concept may change. If there are major shifts in the design concept, design and manufacturing should be required jointly to manage the product and process designs. Also, this cooperation should begin in the early stages of the design process. Interactions between design and manufacturing personnel should involve the transfer of rich information through face-to-face meetings and frequent phone conversations. Barriers in the organizational structure can also be reduced by appointing formal liaison personnel selected from experienced and respected senior staff.

REFERENCES

Andreasen, M. M. "Design Strategy." *Proceedings of the International Conference on Engineering Design,* Boston, 1987.

Beitz, W. "General Approach of Systematic Design." *Proceedings of the International Conference on Engineering Design,* Boston, 1987.

Burns, T., and G. M. Stalker. *The Management of Innovation.* London: Tavistock, 1961.

Campbell, R. B. *Klein Symposium on The Management of Technology,* University Park, Penn.: The Pennsylvania State University, August 24, 1990.

Daft, R. L., and N. B. Macintosh. "A Tentative Exploration into the Amount and Equivocality of Information Process in Organizational Work Units." *Administrative Science Quarterly* 26, 1981, 207–24.

———— and R. H. Lengel. "Information Richness: A New Approach to Managerial Behavior and Organization Design." In *Research in Organizational Behavior,* eds. B. M. Staw and L. L. Cummings. Greenwich, Conn.: JAI Press, 1984, Vol. 6, 191–233.

———— and R. H. Lengel. "Organizational Information Requirements, Media Richness and Structural Design." *Management Science* 32(5), 1986, 554–471.

Ebert, R. J., E. A. Slusher, and K. M. Ragsdell. "Information Flows in Product Engineering Design Productivity." In *Engineering Management: Theory and Applications,* eds. D. J. Leech, J. Middleton, and G. N. Pande. Redruth, Cornwall, England: M. Jackson & Son, 1986, 329–336.

Ehrlenspiel, K. "Reduction of Product Costs in West Germany." *Proceedings of the International Conference on Engineering Design,* Boston, 1987.

Galbraith, J. R. *Organizational Design.* Reading, Mass: Addison-Wesley, 1977.

Gill, H. "Design for Manufacture—A Case Study." *Proceedings of the International Conference on Engineering Design,* Boston, 1987.

Holland, W. E., B. A. Stead, and R. C. Leibrock. "Information Channel/Source Selection as a Correlate of Technical Uncertainty in a Research and Development Organization." *IEEE Transactions on Engineering Management,* EM-23, 1976, 163–167.

Katz, R., and M. Tushman. "Communication Patterns, Project Performance, and Task Characteristics: An Empirical Evaluation and Integration in an R&D Setting." *Organizational Behavior and Human Performance* 21(2), 1979, 139–62.

Lawrence, P. R., and J. W. Lorsch. *Organization and Environment: Managing Differentiation and Integration.* Homewood, Ill.: Richard D. Irwin, 1967.

Middendorf, W. H., and C. L. Wang. "A Standardized Data Base of Product Performance." *Proceedings of the International Conference on Engineering Design,* Boston, 1987.

Morley, I. E., and S. Pugh. "The Organization of Design: An Interdisciplinary Approach to the Study of People, Process and Contexts." *Proceedings of the International Conference on Engineering Design,* Boston, 1987.

Naveiro, R. M. "Product Morphology and Production Automation." *Proceedings of the International Conference on Engineering Design,* Boston, 1987.

Nevill, G. E., and G. H. Paul. "Automated Design of Mechanical Structure Configurations." *Proceedings of the International Conference on Engineering Design,* Boston, 1987.

Primrose, D. G. U. "ND Technovision: A European CAD/CAM System for Mechanical Applications." *Proceedings of the International Conference on Engineering Design,* Boston, 1987.

Pugh, S. "Design Decision—How to Succeed and Know Why." DES 81, Birmingham, England, 1981.

———. "Total Design, Partial Design: A Reconciliation." *Proceedings of the International Conference on Engineering Design,* Boston, 1987.

Sato, N. "Quality Function Expansion and Reliability." *Standardization and Quality Control,* March 1983, 19–27.

Slusher, E. A., and K. J. Roering. "Designing a Scientific and Technical Information System: Behavioral Dimensions and Administrative Decisions." *Urban Systems* 3, 1978, 201–210.

Tannenbaum, A. S. *Control in Organizations.* New York: McGraw-Hill, 1968.

Thompson, J. D. *Organizations in Action.* New York: McGraw-Hill, 1967.

Tushman, M. L. "Technical Communication in R & D Laboratories: The Impact of Project Work Characteristics." *Academy of Management Journal* 21, 1978, 624–645.

——— and D. A. Nadler. "Information Processing as an Integrating Concept in Organizational Design." *Academy of Management Review* 3, 1978, 613–624.

Varljen, L. C. *Klein Symposium on The Management of Technology,* University Park, Penn.: The Pennsylvania State University, August 24, 1990.

Wallace, K. M., and C. Hales. "Detailed Analysis of An Engineering Design Project." *Proceedings of the International Conference on Engineering Design,* Boston, 1987.

Weick, K. E. *The Social Psychology of Organizing.* 2nd ed. Reading, Mass.: Addison-Wesley, 1979.

9

MANAGING DFM: LEARNING TO COORDINATE PRODUCT AND PROCESS DESIGN

PAUL S. ADLER

As firms experience greater pressure on new product time to market, the management of the design-manufacturing interface becomes a more important competitive variable. Too often, designs are "thrown over the wall" to manufacturing, only to discover that numerous engineering design changes are needed to reduce costs and improve quality. One electronics company I have studied analyzed their engineering change activity and found that engineering changes accounted for at least 20% of the business's total overhead costs, that 80% of these changes were avoidable, and that the average cost of each of these avoidable changes was $54,000.

A growing number of companies are thus attempting to remodel their design-manufacturing interface to ensure greater design for manufacturability (DFM) (Dean and Susman, 1989; Krubasik, 1988; Whitney, 1988.) To this end they are experimenting with design teams, design rules, transition teams, CAD/CAM integration, and a variety of other coordination mechanisms. Each of these mechanisms has its fervent partisans in industry and in the academic and consulting communities. Two questions are thus posed. The first question is for project managers: How should a project manager select the mix of mechanisms most appropriate for the specific projects to be managed? The second is for managers with broader responsibilities: How can they assure that their organizations learn better coordination approaches over time?

The available organizational theory does not provide a very useful answer to the first question. The theory developed to date is useful for classifying various types of interdependence in interdepartmental relations and for identifying the coordination mechanisms appropriate for each type of interdependence (see the review by McCann and Galbraith, 1981). But this theoretical framework misses an important dimension of the product development process: it does not acknowledge that at different phases of a product development project, the participating departments experience different degrees of interdependence and coordinate via different coordination mechanisms. Their underlying interdependence is therefore not a constant over the duration of the project. A useful theory of managing for manufacturability should guide the selection of coordination mechanisms in each of the different phases of the project.

As for the second question, how to assure that the organization improves its coordination capabilities over time, here the available organization theory offers more fruitful suggestions, but these suggestions tend to be scattered across distinct research subfields. These insights need to be integrated into a framework useful for guiding the organizational learning process.

This chapter proposes some key elements of a response to both these questions. The following section introduces a new typology of the different coordination mechanisms available in the different project phases. The next section identifies two factors that should, normatively speaking, inform the choice of coordination mechanisms within and across the phases. Finally, I identify the "learning paths" that lead to enhanced DFM capability and the factors that facilitate or impede an organization's learning efforts. An Appendix describes the research base of my analysis.

A TYPOLOGY OF COORDINATION MECHANISMS

Organization theory has identified four generic coordination mechanisms: standards and rules, plans and schedules, mutual adjustment, and teams (March and Simon, 1958; Thompson, 1967; Van de Ven, Delbecq, and Koenig, 1976). Each of these can be used to great effect to coordinate product design and process design and thereby ensure the manufacturability of product designs:

1. *Standardization or rules.* If DFM standards, guidelines, and rules are comprehensive and accurate enough, they can in some circumstances enable the product development and process design staffs to operate rather independently of each other. If manufacturing can be assured that whatever the specifics of the product design, it will satisfy some set of producibility requirements, and if the standards are tight enough, manufacturing can begin many preparatory activities even in the absence of a specific design.
2. *Plans and schedules.* There are many situations in which DFM guidelines do not guarantee a fully producible product, and it is necessary to schedule a sign-off to give manufacturing a chance to double-check the producibility of the specifications.
3. *Mutual adjustment.* In some other situations, even the combination of DFM rules and sign-off authority does not suffice, and the project manager needs to organize a series of design reviews during the design process to enable the product and process designs to be adjusted to each other.
4. *The team.* In complex cases, it is advisable to supplement all the other mechanisms with the formation of an interdepartmental team to ensure the real-time joint optimization of product and process design choices.

In the new product development process, distinct variants of these generic coordination mechanisms can be applied in each of the phases of the product development. For the purposes of the present discussion, I will distinguish three phases: (a) pre-project coordination during the activities that precede the initiation of a given development project, (b) design-phase coordination during the phase dominated

Table 9.1. A Typology of Coordination Mechanisms

	Pre-project phase	Design phase	Manufacturing phase
Non-coordination	Anarchy	Over-the-wall	Work-arounds
Standards	Compatibility standards	Design rules or tacit knowledge	Manufacturing flexibility
Plans and schedules	Development schedules	Sign-offs	Producibility exceptions resolution plan
Mutual adjustment	Coordination committee	Design reviews	Producibility engineering changes
Teams	Joint development	Joint design team	Transition team

by product and process definition, and (c) manufacturing-phase coordination after the release of detailed specifications to the manufacturing department. The output of the pre-project phase is a set of design and manufacturing capabilities; the output of the design phase is a set of product and process specifications and the associated drawings; and the output of the manufacturing phase is shippable product. We could refine this three-phase characterization to distinguish between conceptual design and detailed design, and between pilot production and mature production. In practice, coordination often takes different forms within these sub-phases. The discussion below will make such finer distinctions when they are appropriate, but for the sake of expositional simplicity I will leave the overall conceptual framework in this three-phase form.

In each phase, we can distinguish modes of interaction based on the four generic coordination mechanisms of standards, schedules, mutual adaptation, and teams, and we can contrast these interaction modes with the "base case" alternative of ignoring coordination requirements altogether. The typology is summarized in Table 9.1, and the following subsections discuss each phase in turn.

Pre-Project Coordination

Design and manufacturing can sometimes satisfy much of their overall coordination requirement prior to any specific product development project. In one company I have studied, this is called "filling the pizza bins." The commercial pizza parlor's personnel fill a set of pizza bins so that the making of any specific pizza does not have to wait for the preparation of the ingredients. Similarly, the new product development project should be able to draw on a set of proven and compatible product and process technologies, rather than having to invent the technologies required to realize its project objectives (see Hayes, Wheelwright, and Clark, 1988, chapter 10).

CAD/CAM provides a nice example of such pizza bins, and the various approaches to CAD/CAM development illustrate the various, more and less interactive approaches taken to filling them. Other key pre-project activities that require coordinated effort by design and manufacturing staffs include formulating functional strategies for the two functional departments, developing DFM skills, setting producibility standards, and creating approved parts databases. A similar range of interdepartmental interaction modes can be identified in each of these activities, but to simplify the exposition, I will focus on CAD/CAM development.

The base case of non-coordination of CAD and CAM development is not all that uncommon. Indeed, one aerospace company I have studied (Company B in the Appendix) had a deliberate policy of not attempting to coordinate CAD and CAM development efforts. In an approach that could be called "energetic anarchy," the company encouraged its functional departments to plunge ahead into whatever automation efforts passed a rather generous set of investment criteria and without any constraints on system compatibility. They ended up with 23 different and incompatible computer-based systems in different departments. This strategy was nevertheless considered successful, since they found that the accumulated automation experience and skill-base outweighed the inconvenience and cost of having to reprogram or replace some systems when they decided to integrate them. More often, however, the absence of strategic coordination was less well motivated than at Company B. At another company I have studied (L), the manufacturing manager put it simply: "We never did have and still don't have a CAD or a CAM strategy. So how could we coordinate them?" Project managers exercised considerable power in the organization, and no single project had any incentive to invest in systems that would pay off only over several projects.

For organizations that do want to coordinate CAD and CAM development, several coordination mechanisms are available. At the team end of the spectrum, design and manufacturing engineers can be brought into a new CAD/CAM department jointly to develop and implement a long-range CAD/CAM strategy (Company C). A less interactive model that still allows for some degree of mutual adaptation is a CAD/CAM committee that regularly brings together staff from the different functions to coordinate their activities (Companies F and G).

At the other end of the spectrum of pre-project coordination mechanisms, compatibility standards allow the organization to minimize the direct interaction of the functions in the elaboration of CAD/CAM strategies and still maintain a certain degree of consistency. Some organizations do not develop comprehensive plans for CAD/CAM, but instead the design and manufacturing functions are free to use any systems they choose with the proviso that all the systems must be able to communicate with a central product-definition database (Company I).

Between minimal coordination by compatibility standards and coordination by teamwork or committee lies an intermediate type of CAD/CAM strategy, coordination via schedules and plans. With the help of CAD and CAM specialists in design and manufacturing, a corporate task force can put together a schedule for the development and integration of CAD and CAM. This mechanism requires less ongoing coordination effort but creates no formal authority over the execution of the plan and no forum for resolving the compatibility issues that emerge in implementing the plan (Company A).

Design-Phase Coordination

Under competitive pressure, many firms are seeking closer coordination between design and manufacturing functions during the design phase. But it is important to note that the "over-the-wall" base case is not a caricature. Many companies still have no mechanism whatsoever for discussing producibility issues during the design phase; when designs are released to manufacturing, manufacturing has no opportunity to raise objections, and manufacturing's official mission is "to make whatever comes over the wall" (Companies G and J).

Standards can be a powerful coordination mechanism in the design phase. If the organization develops an explicit characterization of its manufacturing capabilities in the form of producibility design rules, these rules can be used by design engineers to control the producibility of their designs without any direct interaction with manufacturing staff. In the design of printed circuit boards (PCBs), rules specify parameters such as the width and spacing of the lines that can be reliably printed onto the PCB and the pad sizes required to solder components to the board effectively. The effectiveness of such standards is evidenced by the experience of one firm (Company K), where the proportion of board fabrication specifications that were producible the first time increased from 40% to 95% over a two-year time span due to the development and implementation of such design rules.

A second form of coordination by standards relies on design engineers' tacit knowledge of the manufacturing constraints, rather than explicit knowledge coded into rules. When design engineers accumulate experience in manufacturing or accumulate an understanding of producibility constraints through their project experience, they can anticipate and avoid producibility difficulties without direct interaction with manufacturing staff. Too few organizations (only two in my sample—E and H) have systematic job rotation or internship programs to encourage the development of these skills.

A somewhat more interactive form of design-phase coordination is the sign-off procedure, through which manufacturing signals that it accepts responsibility for making a product to the design specifications. This procedure gives manufacturing the right to veto the specifications as infeasible or to refuse to accept responsibility because some of the required documentation is lacking; but this procedure does not create a forum in which product-process fit issues can be negotiated in any detail. Many organizations have such a sign-off procedure, but sometimes the manufacturing organization lacks the power to exercise its veto (Company G).

An even more interactive coordination procedure is the producibility design review. Because of competitive pressure to improve manufacturability, it is increasingly common for organizations to conduct reviews to ensure that producibility considerations are being respected. In-process reviews allow for revisions to be incorporated before too much effort has been expended to optimize the design from a performance point of view.

At the most interactive end of the spectrum, we find some companies experimenting with a product-process team approach (in our sample, E, H, O, P, R, and U; for a general discussion of this approach, see Dean and Susman, 1989). These teams bring manufacturing engineers into the design process both to begin developing process designs as early as possible and to offer product designers informal advice on how to enhance the producibility of their emerging designs.

Manufacturing-Phase Coordination

The need for coordination usually continues well into the manufacturing phase. Absent a formal mechanism for assuring this coordination, when the manufacturing organization encounters serious producibility problems it will often take matters into its own hands and make covert changes to the product design— "work-arounds." There are, of course, several alternatives to work-arounds; as in the other phases, there is a spectrum of coordination mechanisms, from those based more on standards to those requiring more direct collaboration.

At the standards end of the spectrum, investments in manufacturing flexibility can assure a de facto coordination between product and process design. In the pre-CAD/CAM days, the only way to achieve high levels of flexibility in manufacturing was to avoid the use of specialized equipment and to rely on general-purpose equipment—and thereby to incur higher average operating costs. But CAD/CAM and other engineering innovations have mitigated this trade-off. For example, newer CAM systems for PCB component insertion often have storage capacities for a larger number of different types of components and their computer controls enable them to alternate between board designs at minimal costs. By enlarging the "envelope" of product designs that can be handled without new setups, manufacturing flexibility effectively facilitates the coordination of product and process design.

Manufacturing-phase coordination sometimes requires changes to the product design. When these changes can be anticipated, for example when they have been identified at the final design review, the organization can develop a detailed schedule for resolving "DFM exceptions."

Engineering changes (ECs) represent a common form of mutual adaptation: in a frequently encountered scenario, design "throws the drawings over the wall" to manufacturing, and manufacturing sends back a list of changes that have to be made in order to make the design producible. (ECs are also the way the organization coordinates the implementation of minor changes requested by customers, proposed by marketing, or occasioned by design flaws or vendor changes.) In the case of aircraft hydraulic tubing, before the introduction of CAD/CAM, first-time-fit ratios (the proportion of tubes requiring no adjustment going from the third mock-up to the first regular production aircraft) averaged between 10% and 20%, necessitating a huge flow of ECs. Not only does CAD/CAM help avoid many of the errors that occasion ECs, it also helps manage the EC cycle more efficiently by ensuring faster processing of design changes.

Under pressure to ensure a higher-quality product-process fit, some firms (in our sample, Companies E, H, O, P, R and U) have established "transition teams." In this approach, some design engineers move with the design into manufacturing on temporary assignment, so as to make themselves available on a full-time basis for whatever design revisions are required. This mechanism helps deal with a common problem: design personnel who have moved on to the next product design project after the last one was released to manufacturing are typically reluctant to give ECs for the previous project as high a scheduling priority as their new product development activities. This rotation is also a way of developing manufacturing understanding on the part of design engineers.

SELECTING THE RIGHT COORDINATION MECHANISMS

While the more interactive mechanisms have many proponents, it is important to recall that they are very burdensome in meeting time. If coordination can be assured through less interactive means, they are to be preferred. In the other dimension of my typology, it is clear that earlier coordination is preferable to later—the more effective the earlier coordination effort, the less interactive, and therefore the less expensive, will need to be the coordination effort in later phases, and the shorter will be the overall time to market. But how is the project manager to select the optimal set of coordination mechanisms? This section identifies two criteria that should be used if the organization is to select the coordination mechanism that is optimal for the task at hand. I provisionally will bracket the political and cultural factors that may impede the organization's recognition or implementation of this "optimal" organizational design: I return to these factors in the following section.

The task to be accomplished by design-manufacturing coordination mechanisms is that of ensuring the DFM fit between product and process characteristics. Depending on the degree of uncertainty of this fit, different coordination mechanisms are needed. Following a long line of research in organization theory (starting with Perrow, 1967: and extending to Van de Ven and Delbecq, 1974; Van de Ven, Delbecq, and Koenig, 1976; Tushman, 1979; Daft and Macintosh, 1981; Withey, Daft, and Cooper, 1983; Fry and Slocum, 1984; Victor and Blackburn, 1987), I propose to conceptualize product-process fit uncertainty in two dimensions: task exceptions (the number of factors observed as exceptions to what is known) and search difficulty (the difficulty of resolving each of these exceptions). My research suggests that these two dimensions are indeed the key determinants of the appropriateness of a given coordination mechanism to a given DFM problem.

Let us take first the task exceptions dimension. The set of DFM problems presented by any given component can present a greater or smaller number of exceptions with respect to the experience base of the organization—I call this characteristic the novelty of the DFM problem. A greater degree of novelty creates uncertainty by making the choice of product design parameters more sensitive to the choice of process design parameters or vice versa. Greater novelty therefore calls for more intensive use of the available information and thus should lead the organization to use more interactive coordination mechanisms.

Some examples will make clearer the nature of this novelty/interaction link. Take first the design phase. The novelty/interaction link has been analyzed carefully by one of the firms I have studied (Company E). They have developed an explicit set of criteria for deciding how much interaction a given project will need. They distinguished four levels of interaction: (1) rely on standards, then use manufacturing prototypes to resolve residual fit issues; (2) conduct a meeting with manufacturing staff early in the design phase to set some general parameters, then rely on the sign-off to ensure that they have been respected; (3) designate liaison people to conduct occasional in-progress design reviews; and (4) implement a full product-process design team. The choice of interaction level is based on a number of factors, most prominently whether the product and process technologies were (1) proven carry-overs from earlier projects, (2) minor refinements, (3) major changes, or (4) unproven new approaches.

Turning to the degree of interaction during the manufacturing phase, we can easily see that if the product-process fit issues have all been experienced in previous projects, the flexibility of the existing manufacturing procedures will be able to cope with the new product release and there will no need to redefine specifications as a result of a prototype experience. If, however, the product-process fit issues embody exceptions with respect to prior experience, the organization will need to plan for some design changes. If the number of exceptions is high, advanced planning will not be able to anticipate them, and the manufacturing function will subsequently need to propose fit-enhancing changes in the form of ECs. If the novelty is so high that the organization expects a very large number of ECs, then a transition team that brings some design engineers into manufacturing will be very useful.

The same novelty/interaction link works in the pre-project phase. Using the example of CAD/CAM again, I have found that as long as innovations in the design and manufacturing technology of PCBs are incremental, effective coordination of CAD and CAM strategies can be assured by the rule that all CAD and CAM systems had to be able to communicate with the product definition database. But if the design parameters and the manufacturing process are changed substantially—for example, with the shift from the traditional through-hole technology and into surface-mount technology—CAD and CAM strategy coordination can only be assured by the formation of a core CAD/CAM group that could manage all the fit issues in a timely manner.

The second dimension of DFM fit uncertainty is analyzability— the difficulty of the search for the answer to the given DFM problem. Some fit problems can be resolved very easily through recourse to well-established know-how; but other problems stretch current DFM know-how and require a more taxing problem-solving process. Lower analyzability therefore calls for the creation of new product-process fit information, in particular by passing from the very abstract and generic characterizations of products and processes that guide the pre-project capabilities development activity, to the less abstract and more specific characterizations that emerge from the design phase in the form of drawings and specifications, to the very concrete characterizations of product and process that are created with the manufacturing output. While high novelty of the DFM fit problems creates the need for more interaction during a given phase, low analyzability forces the project to postpone the resolution of some of those fit problems to later phases, where new information will be created.

The power of pre-project coordination in cases of very high analyzability was visible at a firm offering design and foundry services for application specific integrated circuits (ASIC) development (Company N). An extensive pre-project effort conducted jointly by design and manufacturing people had enabled them to code all the relevant manufacturability knowledge into their CAD database. As a result, coordination during the product development project itself could be assured with minimal effort, greatly accelerating the cycle: during the design phase coordination could rely exclusively on design rules, and during the manufacturing phase coordination

relied exclusively on manufacturing flexibility. In this case, all the requisite manufacturability knowledge can be coded because it extends not much further than a
characterization of the line and space widths that can be reliably reproduced—the key
design rules can fit on a single page of text. The analyzability of the fit problems in
this case is thus almost total. Even for designs that involve thousands or millions of
exceptions with respect to the specific details of product and process in previous
designs, all the coordination effort can be pushed into the pre-project phase.

Engine manufacturing (Company O) represents the other extreme. The nature of
the mechanical engineering and metal forming tasks make exclusive reliance on design rules impossible. Despite their best efforts to characterize their manufacturing
capabilities, numerous product design changes are made after release to manufacturing. Given their high volumes, intense cost competition, and severe quality requirements, engine designs are continually refined for several years into mature manufacturing. In order to ensure the aggressive pursuit of these changes and the timely
processing of the resulting ECs, Company O had assigned the design engineers to
remain responsible for their product for its entire life, effectively acting not just as a
transition team but as a "life-cycle" team.

The selection of optimal coordination mechanisms thus depends on two dimensions: the choice in the interaction dimension—between standards, plans, mutual adaptation, and teams—should be a function of the degree of novelty of the product-
process fit problem, and the choice in the temporal dimension—between pre-project,
design, and manufacturing phases—should be a function of the analyzability of that
fit problem. Any given development effort will involve more than one product-process
fit problem, and these different problems will typically evidence different degrees of
novelty and analyzability. So the optimal coordination approach for the project will
involve a portfolio of mechanisms, the mix being determined by the relative importance of the different types of fit problems.

IMPROVING DFM OVER TIME

The preceding sections have identified several coordination mechanisms, arrayed them
in a typology, and identified two criteria that can serve to identify the optimal mechanism for a given DFM problem. But in practice, the organization does not begin
with a blank slate—engineers and managers are already accustomed to using some
set of mechanisms. So even if theory allows us to identify an optimal mechanism, a
key problem remains to be addressed: How can management orchestrate the process
of organizational learning needed to ensure that the organization progressively improves its coordination capability? This section addresses the learning process issues
through the exposition of three models: a model of the learning process, a model of
the factors that facilitate or inhibit learning, and a model of the cultural dimensions
of DFM learning.

The Learning Process

It is useful to think of this learning process in the form of the following causal model.
In any given situation, there is a certain (more or less exogenously given) degree and

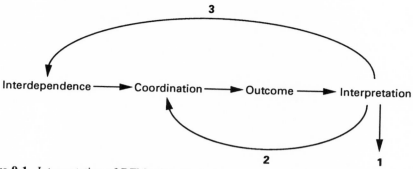

Figure 9.1. Interpretation of DFM-related outcomes.

form of interdependence between product design and process design. As we have seen, this interdependence varies across industries, businesses, and projects. To assure maximum effectiveness, the interdepartmental coordination mechanisms should be designed to manage this interdependence.

In reality, the coordination mechanisms in place in a given organization at a given time reflect a wide range of other constraints—the established coordination practices reflect the history of the organization. Depending on how well these mechanisms are capable of dealing with the real interdependence, the outcome will be a better or worse DFM performance, as reflected in the characteristics of the product (such as cost, quality, and time to market) and the market's reaction to these characteristics. This outcome is then interpreted by the organization. While different interpretations might be adopted by different actors within the organization, the dominant coalition's interpretation can be expected to prevail in general.

The resulting interpretation is one of three types, each of which suggests a certain course of action: (1) the outcome is seen as acceptable, or performance problems are not attributed to DFM, and therefore the organization continues with business as usual (this same conclusion could also be reached without any active interpretation at all); (2) the outcome is unsatisfactory and the problems are attributed to DFM, so there is some pressure to refine the implementation of the current coordination mechanisms; (3) the DFM outcome is seen as radically deficient and therefore a whole new approach is seen as necessary. Figure 9.1 depicts these three paths (see Figure 9.1).

Figure 9.1 depicts the third path as involving the development of a new interpretation of the underlying interdependence. A variant of this path might also be possible: in some organizations new coordination mechanisms are instituted by top management fiat (''You will use joint product-process design teams!''), and it is through the lived experience of using these new coordination mechanisms that the participants come to form a new understanding of the underlying interdependencies.

Factors Facilitating and Impeding Learning

The focal question of this section can now be reformulated: What leads an organization to adopt a given organizational learning path rather than another? The model

I propose for thinking about this question extends those of Louis and Sutton (1991) and Gersick and Hackman (1990).

The choice of a less aggressive path—path 1 over paths 2 or 3, or path 2 over path 3—may reflect a truly satisfactory DFM level, but might also reflect several other factors:

Inertia. One of the most powerful inhibiting factors is the least visible—simple force of habit. Even when joint design teams are formed, members sometimes hold tight to their established ways of doing things, to "tried and true" methods of working.

Lack of resources. Change can be blocked by scarcity of resources such as time, people, and specialist skills with which to design or implement new coordination mechanisms. Many DFM improvement efforts are constrained because their design and/or manufacturing engineers simply do not know how to perform a DFM analysis. Many design engineers have had neither manufacturing experience nor DFM training. In some cases, manufacturing staff know how to analyze a design proposal and identify its producibility problems through prototyping, but they find it very difficult to predict producibility problems based on the sketchy information that is available earlier in the design process. Design rules cannot be developed without devoting time and skills to the quantitative characterization of the limits of manufacturing capabilities.

The political costs of change. When design and manufacturing managers are engaged in turf battles, as is often the case, DFM initiatives become hostage to the established power balance. Part of the problem is simply that the cost of change is often much more visible than the benefits. These political tensions are not only interfunctional; middle-level managers often resist giving up decision-making power to their subordinates on the development team.

Norms/values/schemas. Design engineers often see themselves as smarter than manufacturing engineers and therefore behave like prima donnas. They are sometimes recruited from more prestigious schools, and top management often signals in numerous ways that they have a higher status in the organization.

Factors that appear to be particularly powerful in encouraging a firm to adopt a more aggressive learning path include the following.

Business crises. Severe economic pressure—in the form of financial results so poor as to endanger the survival of the business—seems to be a necessary condition for aggressive DFM improvement. Not one of the organizations that I have studied has made substantial improvements in their DFM performance without a business crisis.

Demands from above. While a business crisis appears to be a necessary condition, it is not a sufficient condition—top management has to lead the organization to see poor DFM performance as the cause of the crisis. During the 1970s, Company R, a communications equipment manufacturer, lost much of its market share to Japanese competitors. Top management interpreted this loss of market share as being due to unfair competition. This assumption persisted until, in the course of preparing an appeal for trade protection to the federal government, top managers were forced to compile and examine more objective data. This effort led them to see the real source of their problems in their products' quality and their insufficient focus on DFM. Top management then mobilized the sense of crisis into an organization-wide change in attitudes, priorities, and new product development practices.

Technological pressure. I have encountered one case in which a real, albeit temporary DFM improvement occurred without a business crisis. In this case, aircraft Company B was introducing a new material, titanium, and the design engineers knew that their products risked catastrophe if they did not collaborate with manufacturing engineers to ensure DFM.

Environmental pressures. Some defense contractors (such as Company S) have been forced to undertake substantial DFM improvement efforts in response to new Department of Defense requirements. The DOD has recently begun requiring that bids for development contracts include a producibility plan, and this has created pressure on the contractors to change their practices.

Finally, there is a set of factors that are not strong enough by themselves to overcome resistance but do lend support to those in the organization pushing to adopt a more aggressive posture. The presence of a lower-level champion can be an important facilitating factor, although I have not found any cases in which such a champion drove significant change without pressure from higher-level management and a context of crisis. The visibility of organizations with more effective approaches to coordination sometimes can serve as a stimulus for change. A report from the Institute For Defense Analyses (1988) seems to have played this role for a number of defense contractors. Some organizations have found that the momentum for DFM improvement is impeded by the lack of metrics for DFM: when DFM performance can be assessed in objective rather than subjective terms, proponents of DFM can make their concerns more salient.

DFM as Cultural Change

The most difficult challenges of improving DFM are often at the cultural level. This cultural challenge can be analyzed at several levels of visibility: artifacts, values, and basic assumptions (Schein, 1984). At each level, many firms confront the cultural legacy of a past in which product design and manufacturing were separated by a chasm.

Artifacts. In many firms, design and manufacturing engineers are not only not at the same average pay levels, but they are not even on the same pay curves. A multiplicity of other organizational artifacts communicate the same message of inequality: amount of office space, time to participate in professional activities, etc.

As organizations "get serious" about DFM, these artifacts of organizational life often change and new ones emerge:

- New reward and promotion criteria are introduced
- DFM assessment methodologies are institutionalized

- "War rooms" are designed where product-process design teams get together for weekly meetings and post the work-in-progress

Values. As firms become more committed to DFM, they typically experience shifts towards values listed below:

- Greater trust between the functions
- A priority on getting the design right the first time it is released—this leads to more early iterations
- A personal identification with the enterprise's objectives rather than with parochial departmental objectives
- A focus on predicting manufacturing problems rather than reacting to them
- Greater accountability of the product-process design team rather than a narrow view of one's responsibilities
- Design for simplicity rather than sophistication

Basic assumptions. The most difficult part of the shift towards greater emphasis on DFM is often the change it requires in underlying assumptions about the way the world works. I have already mentioned the importance of top management's interpretation of the source of the business's competitive difficulties; but the assumptions made by the engineers themselves also play a key role. Key impediments at the level of assumptions include the following:

- The value of extra time spent very early in the development cycle—when people are eager to dive into the detailed design task—is sometimes hard to accept,
- Some design engineers believe that the business's interests are best served when their creative autonomy is least constrained and they thus cannot understand why they are being asked to take on the task of trading off product parameters and process design choices,
- Some design engineers assume that problems with the product design would most appropriately be solved by adding a part to it—rather than by eliminating a part,
- Some design engineers and some manufacturing engineers assume that several iterations through production prototype are always required—the idea of getting it right the first time through more extensive up-front analysis seems utopian,
- Engineers and managers in both departments sometimes feel that they are under so much pressure to meet schedules that they have no time to develop new ways of doing things and no time to gather and incorporate input from lower levels or from other departments.

CONCLUSIONS

This chapter has made three contributions to our understanding of the management challenge of DFM: defining a typology of coordination mechanisms; identifying the key criteria that should inform the selection of the optimal mechanism for a given type of DFM task; and characterizing the DFM learning process. In conclusion, it is appropriate to highlight some of the limitations of these contributions and to suggest some avenues for future research.

With respect to the first contribution, the typology, future research should lift the assumption of a preexisting departmental specialization of product design and manufacturing. As a result of their experience with joint teams and transition teams, some companies have created a single staff of "product engineers" who assure both product design and manufacturing engineering functions.

With respect to the second contribution, both the testing and the practical utility of the approach developed here depend entirely on our ability to construct valid and reliable measures of DFM novelty and analyzability. Daft and Macintosh (1981) show that this can be done for the uncertainty and equivocality of individual tasks. Future research will need to confirm that it can be done for coordination tasks.

Finally, the learning process model needs further refinement. Future research should lift the assumption that the design- manufacturing interdependence is an exogenous given. Firms can change the nature of this interdependence, at least within certain limits. Some companies have drawn considerable benefit from the use of "approved parts" databases that constrain designers to use components whose manufacturing characteristics have already been thoroughly documented—they deliberately limit novelty. Other companies invest considerable efforts to characterize their manufacturing capabilities so as to reduce the analyzability problem. These kinds of initiatives shift our focus from the operational level that has been the concern of this chapter to the strategic level. Interfunctional coordination for DFM is certainly strategically important enough to make such research a high priority.

ACKNOWLEDGMENT

This research was supported by McKinsey & Co. and by the Stanford Institute for Manufacturing and Automation. Managers and engineers at the companies surveyed graciously provided not only data but also their valuable insights. They must unfortunately remain anonymous. Duane Helleloid, Elaine Rothman, and Reuven Regev provided extensive research assistance.

REFERENCES

Daft, R. L., and N. B. Macintosh. "A Tentative Exploration into the Amount and Equivocality of Information Processing in Organizational Work Units." *Administrative Science Quarterly* 26(2), 1981, 207–224.

Dean, J. W., and G. I. Susman. "Organizing for Manufacturable Design." *Harvard Business Review* 67(1), January-February 1989, 28–36.

Fry, L. W., and J. W. Slocum. "Technology, Structure and Work Group Effectiveness: A Test of a Contingency Model." *Academy of Management Journal* 27, 1984, 221–246.

Gersick, C. J. G., and J. R. Hackman. "Habitual Routines in Task-Performing Groups." *Organizational Behavior and Human Decision Processes* 47(1), October 1990, 65–97.

Hayes, R. H., S. C. Wheelwright, and K. B. Clark. *Dynamic Manufacturing.* New York: Free Press, 1988.

Institute for Defense Analyses. The Role of Concurrent Engineering in Weapons Systems Acquisition, IDA report R-338, Alexandria, Va: 1988.

Krubasik, E. G. "Customize Your Product Development." *Harvard Business Review* 66(6), November-December 1988, 46–53.

Louis, M. R., and R. I. Sutton. "Switching Cognitive Gears: From Habits of Mind to Active Thinking." *Human Relations* 44(1), January 1991, 55–76.

March, J. G., and H. A. Simon. *Organizations.* New York: Wiley, 1958.

McCann, J. E., and J. R. Galbraith. "Interdepartmental Relations." In *Handbook of Organizational Design,* eds. P. C. Nystrom and W. H. Starbuck. New York: Oxford University Press 5(2), 1981, 60–84.

Perrow, C. "A Framework for the Comparative Analysis of Organizations." *American Sociological Review* 32, 1967, 194–208.

Schein, E. H. "Coming to an Awareness of Organizational Culture." *Sloan Management Review* 25(2), Winter 1984, 3–16.

Thompson, J. D. *Organizations in Action.* New York: McGraw-Hill, 1967.

Tushman, M. L. "Work Characteristics and Subunit Communication Structure: A Contingency Analysis." *Administrative Science Quarterly* 24, 1979, 82–98.

Van de Ven, A. H., and A. L. Delbecq. "A Task Contingent Model of Work-unit Structure." *Administrative Science Quarterly* 19, 1974, 183–197.

———, A. L. Delbecq, and R. Koenig, Jr. "Determinants of Coordination Modes Within Organizations." *American Sociological Review* 41, 1976, 322–338.

Victor, B., and R. S. Blackburn. "Determinants and Consequences of Task Uncertainty: A Laboratory Study and Field Investigation." *Journal of Management Studies* 24(4), July 1987, 339–404.

Whitney, D. E. "Manufacturing by Design.""Harvard Business Review* 88(4) July-August 1988, 83–91.

Withey, M., R. Daft, and W. H. Cooper. "Measures of Perrow's Work Unit Technology: An Empirical Assessment and New Scale." *Academy of Management Journal* 26, 1983, 45–63.

APPENDIX: RESEARCH BASE

The research on which this analysis is based was inductive in nature: the lessons of firms struggling with the DFM problem served as my key resource. Since I sought to develop a conceptual framework that would have some generality, the field research encompassed several industries.

In a first phase of research, the author and two research assistants studied several organizations designing and building two different types of products—printed circuit boards (PCBs) for electronic assemblies and hydraulic tubing for aircraft. The sample in this phase was selected to highlight cases in which a new cluster of technologies—grouped under the general heading computer-aided design/computer-aided manufacturing (CAD/CAM)—was encouraging firms to reconsider their design-manufacturing interface approach. A total of thirteen organizations was studied in this phase.

In a second phase of research, we visited firms that were known to be particularly aggressive in their pursuit of DFM improvement, whether or not CAD/CAM was a key element of their effort. In this phase, we visited seven organizations in a broader spectrum of industries—semiconductors, electronic assemblies, engines, transportation equipment. Table 9.2 briefly characterizes the sample.

Given our research focus on the coordination of distinct, functionally specialized departments, we sought larger, more mature organizations for whom the differentiation of design and manufacturing functions was a well-institutionalized organizational reality. All the sample organizations were at least 10 years old, and the smallest of them employed 240 people.

The ground rule of the study was that nothing should compromise the anonymity of the participating companies. The contact person was either a general manager whose responsibilities encompassed design and manufacturing or a senior manager directly involved in cross-functional CAD/CAM or DFM efforts. In each organiza-

Table 9.2. Sample Description

Company	Product[a]	Product[b]	Number of interviews
A	Airplanes, hydraulic tubing	Commercial	11
B	Airplanes, hydraulic tubing	Commercial	12
C	Airplanes, hydraulic tubing	Defense	10
D	Airplanes, hydraulic tubing	Defense	7
E	PCBs, low complexity	Engine controllers, low to very high volumes	12
F	PCBs, low complexity	Avionics, very low volumes	10
G	PCBs, medium complexity	Flight simulators, very low volumes	7
H	PCBs, medium complexity	Computer peripherals, medium volumes	14
I	PCBs, high complexity	Mainframe computers, medium volumes	11
J	PCBs, low complexity	Electronic instruments, very low volumes	9
K	PCBs, low to medium complexity	Consumer durables, low to high volumes	7
L	PCBs, medium complexity	Mini-computers, and instruments, low volumes	6
M	PCBs, low to high complexity (depending on customer)	Electronics assembly, low to high volumes	3
N	Semi-conductors	Merchant semiconductors, low volumes	3
O	Engines	Agricultural equipment, high volumes	9
P	Electronic assemblies	Communications switches high volumes	4
Q	Electronic assemblies	Personal computers, high volumes	4
R	Electronic assemblies	Communications products, high volumes	9
S	Mechanical fabrication and assembly	Transportation equipment, low volumes	15
T	Electronic assemblies complexity	Mini-computer equipment for commercial use, medium volumes	12

[a]PCB complexity levels are simple (1 to 2 layers), complex (2 to 6 layers), or very complex (over 6 layers).
[b]PCB volumes are low (less than 500 boards per year), medium (500 to 5,000), high (5,000 to 50,000), or very high (over 50,000).

tion, we interviewed the business unit general manager, managers responsible for DFM and CAD/CAM, functional managers in design and manufacturing, and at least one experienced design engineer and one experienced manufacturing engineer.

A semi-structured interview schedule included both general questions and questions on specific product development projects. The general questions included items on the history of the business unit and its results, the time-line of their experience with DFM and CAD/CAM, the evolution of the organization's skill base, changes in organizational structure, the evolution of business, functional and coordination strategies, and indices of organizational and departmental culture. The items on specific projects were designed to identify similarities and differences in the conduct of one sample of development projects conducted five years earlier and a second sample of projects conducted within the previous year. This part of the discussion focused on the technical and business characteristics of the projects and on the specific design-manufacturing coordination mechanisms used. Each item was included in at least two interview schedules. Semi-structured interviews of between 45 minutes and two hours were conducted separately with each informant.

10
ENGINEERING CHANGE AND MANUFACTURING ENGINEERING DEPLOYMENT IN NEW PRODUCT DEVELOPMENT

PAUL D. COUGHLAN

As manufacturing firms increase their rates of new product introduction, the management of engineering change has become a major and ongoing challenge. IBM, Lockheed, and Northrop, to mention but a few firms, are critically reviewing their approaches to the management of engineering change (Noaker, 1987; Rohan, 1989). Engineering change occurs in product materials, the manufacturing process, or the specifications of the product itself (Hayes and Clark, 1985). In many ways, engineering change is an inevitable consequence of a rapid rate of new product introduction and short product life cycles. Yet, implementation of such change is often disruptive and potentially costly:

> Poorly handled ECOs (engineering change orders) create wasted material as well as increased reject rates . . . One of the most important tasks of management, therefore, is to prevent confusion or mitigate the potentially damaging effects of confusion-causing activities. (Hayes, Wheelwright and Clark, 1988, pp. 182–183)

In controlling or avoiding engineering change, a firm acknowledges the existence of change and attempts to minimize its disruptive effects, while maximizing its positive potential. However, as emphases, control and avoidance differ in their timing. Control treats change as an ongoing phenomenon requiring continuous management. Avoidance suggests a particular cutoff point, usually at the start of routine volume production, after which change is unacceptable. The management challenge is to avoid rather than just control engineering changes, in part through the deployment of technical functions, such as manufacturing engineering, during the new product development process.

This chapter reports on an empirical, exploratory study of engineering change in newly developed products that addressed two specific questions:

1. Is the avoidance of engineering change in newly developed products associated with the way manufacturing engineering staff are deployed during the product development process?
2. Is an association between the avoidance of engineering change and the way these engineering resources are deployed contingent on the development context?

The study addressed these questions through an investigation of 12 products developed between 1983 and 1989 in four divisions of a single company. The company was a major competitor in the electronic equipment industry, with corporate offices located in the United States. An estimated 80% of its current revenue was derived from products less than five years old. Further, designing products right-first-time and design for manufacturability were current strategic issues. A major challenge facing the company was sustaining and improving its aggressive pattern of growth in an increasingly competitive market. The company's vision to the end of the century demanded delivery of products and systems "of the highest quality and reliability, on time, tailored to the varying needs of our customers." These products were to be economical to produce and supported by those who knew the product best.

In brief, the study found that the incidence of engineering change reflected the experience of the manufacturing engineering (ME) staff deployed by the ME manager during the development process, the phasing of their involvement, and their emphasis on manufacturability. The relationship between product performance and ME deployment depended upon the relative newness of the product: newer products were typically less manufacturable and experienced more manufacturability-related change than less-new products. The study concluded that engineering change, which altered the functional specification of a product and was apparent to the customer, was avoidable before the start of routine volume production and consistent with the notion of design right-first-time. In contrast, manufacturability-related change, which was transparent to the customer but apparent to manufacturing, was less avoidable and more consistent with the notion of design right-next-time.

ORIGINS AND IMPACT OF ENGINEERING CHANGES

Firms make engineering changes to effect a design improvement, reduce costs, simplify tooling, facilitate assembly, or rectify an inadvertent or intentional omission at the time of original release (Andrew, 1975). Changes may originate with the product design, manufacturing, test, purchasing, quality, or marketing functions. Engineering changes are implemented during the development phase of a new product, or after the start of routine volume production. Engineering change after manufacturing start up goes against the notion of design-right-first-time, and is most disruptive.

Implementation of engineering change affects more than just the particular parameter in question. Owing to interrelationships among product features, a change to a single feature may precipitate collateral changes to other features (Hauser and Clausing, 1988). Also, an engineering change may apply to parts being developed or manufactured. The cost implications vary accordingly. Preexisting parts may be retained, scrapped or reworked. Parts to be manufactured may be delayed while new engineering details are worked out. In either case, while the change may improve the functional performance of the product, it may also disrupt delivery schedules or cost targets. Many times these costs are hidden in scrap accounts, purchase price variances, field service accounts, or manufacturing inefficiencies (Diprima, 1982).

Yet, however effective, control systems that screen and question engineering changes may not eliminate the negative effects of these changes. Hayes and Clark (1985) found clear evidence of reduced total factor productivity in some plants for up to a year after implementation of an engineering change. They concluded that the average number of engineering changes, their variability from month to month, and the way they were managed all had a measurable effect on total factor productivity. They advocated reducing the number of engineering changes to which a plant must respond in a given period of time and releasing these changes in a controlled, steady fashion rather than in bunches. A firm might accomplish this reduction by exercising discipline on its engineering or marketing people to focus only on the most important changes, or to design right-first-time.

The objective of design right-first-time relates, in part, to the manufacturability of the product. A manufacturable product avoids, on the one hand, production line stoppages, rework costs, and after-sales problems, and, on the other hand, increases safety, quality of workmanship, cost savings, and process compatibility (Wood and Coughlan, 1988). Three common concerns arise in relation to manufacturability: (1) ease of fabrication and handling of individual parts; (2) ease of inspection and test of parts and assemblies; and (3) ease of assembly and associated kitting and handling (Dewhurst and Boothroyd, 1984; Hales, 1987).

These concerns may be addressed by a reduction in the number of parts, the development of foolproof assemblies and a simplified assembly process, use of common components across product families, avoidance of tolerances that exceed process capabilities, and use of modular options (Starr, 1965; Whitney, 1988; Walleigh, 1989).

The product development process provides opportunities to evaluate manufacturability and to make the necessary changes prior to the start of volume production. These opportunities include value analysis, manufacturability assessments, prototype building, and pilot production. Depending on the way in which the development process is managed and the engineering staff deployed, these activities may or may not be carried out, and may or may not be carried out well.

The questions addressed in this study probe the relationships, shown in Figure 10.1, between the incidence of engineering change, the deployment of ME staff, and the development context of the product under development. Before introducing the findings of the study, the dimensions of both ME deployment and of the development context will be discussed and operational definitions presented.

DEPLOYMENT OF MANUFACTURING ENGINEERING STAFF

Manufacturing engineering has responsibility for the development, support, and improvement of products and manufacturing processes (Coughlan, 1989). Accordingly, ME translates product design specifications into simple work instructions and standards, so that production staff will know what they must do to build the product. But before these work instructions can be prepared, ME must determine how the

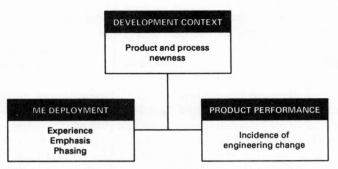

Figure 10.1. Relating engineering change to manufacturing engineering deployment.

product is to be built, what and who will be needed to build it, and what the build schedule will be. After these instructions have been prepared, ME constantly devises ways and means to reduce costs, improve quality, enhance process flexibility or whatever competitive priorities are important to operating success.

ME managers take a number of decisions when deploying their staff on a product development project. These decisions include the phasing of ME staff involvement, when and how much emphasis to place on manufacturability, and the use of ME experience. While these decisions are routine, they are not without risk. Yet, there is little empirical guidance for managers on the nature of the risk, particularly in terms of engineering change, or what they can do to avoid that risk.

Phasing of ME Staff Involvement

Over the course of a product development project, the ME manager assigns staff to support the development effort. ME requires a certain staff size to deliver on its responsibilities during the product development process. Empirical studies suggest that the earlier ME is involved, the better the product performance (Gerstenfeld, 1976; Langowitz, 1989). However, these two studies give no indication of the differences in ME staff requirements among stages of the development process or in different development contexts.

In this study, ME man-hour expenditures over the duration of each project were available for nine products developed in three divisions of the company. Where available, these ME man-hours were grouped by stages in the development process. Accordingly, the phasing of ME staff involvement was measured as the number of man-hours expended during each stage.

Emphasis on Manufacturability

Manufacturability is not implicit in the design of a product. Rather, through developing and applying design rules, and through carrying out manufacturability assessments during the product development process, ME places emphasis on manufacturability. Avoidance of these activities may lead to less manufacturable products through parts mix-ups, missing parts, and test failures in assembling and testing designs with numerous parts (Duck, 1986; Daetz, 1987; Langowitz, 1988; Walleigh, 1989). Even so, the question remains, does ME need to place emphasis on manufacturability in the same way in all development contexts?

To address this question in the study, ME managers were asked to rate on a five-point scale the degree of emphasis placed by ME on 11 dimensions of manufacturability for each product relative to earlier products in its family, or in the previous product family if the product in question was the first of its family. The dimensions of manufacturability included tolerances; the number of parts; component density; the number of free leads; standardization, presentation, orientation, insertion and joining of parts; the need for final assembly adjustments; and self-locating features. This measurement system provided an aggregate "emphasis" score for each product, based on addition of ratings for each dimension of manufacturability. A low score indicated a low degree of emphasis on manufacturability and vice versa.

Phasing of emphasis on manufacturability by ME was measured through identification of the stages of the product development process in which ME placed emphasis on each dimension of manufacturability. This measurement system also provided an aggregate "phasing of emphasis" score for each product. A low score indicated emphasis on manufacturability early in the development process, and vice versa.

Use of ME Experience

Those involved in product development projects engage in a constant process of learning and unlearning, across both levels and functions. The know-how accumulated at the individual level is transferred to other divisions or to subsequent projects within the organization, becoming institutionalized over time (Imai, Nonaka, and Takeuchi, 1985). Through this process of osmosis, the firm transfers and accumulates experience in managing product development (Takeuchi and Nonaka, 1986). Building the technical and managerial abilities of an extended group of people who are assigned according to a careful plan can be used to enhance a firm's ability to plan and execute product development projects (Hayes, Wheelwright, and Clark, 1988).

Similarly, ME staff build up, over time, a unique expertise, which may be lacking in both the development and manufacturing organizations (Szakonyi, 1985). Yet, does the incidence of engineering change reflect the experience of the ME staff assigned by the ME manager during the development process?

In this study, ME staff experience was measured as the extent to which the ME staff involved in a product development project had worked on related and similar projects; had worked with multidisciplinary colleagues; had attained a high level of technical expertise and academic qualification. ME managers rated each dimension of experience on a five-point scale. The sum of the ratings on these dimensions provided an aggregate "experience" score for the ME group involved in each new product development project. A low score indicated a low level of experience, and vice versa.

DEVELOPMENT CONTEXT: PRODUCT AND PROCESS NEWNESS

The context in which products are developed and ME managers operate is defined, in part, by the degree of product and process newness. Newness is one of the most important factors affecting a new product's success or failure (Yoon and Lilien, 1985). Products new to the firm very often require the acquisition of new technological resources, and take the firm into unfamiliar technological territory. An assessment of newness helps to set specifications and targets for individual projects, provides a context for relating concurrent projects, and indicates how the sequence of projects capitalizes on the company's previous investments (Wheelwright and Sasser, 1989).

Writers have conceptualized newness in a number of ways:

- Newness to the firm of customers, product class, needs served, production process, technology, distribution/sales force, advertising/promotion, and competitors (Cooper, 1981)
- Style change, product line extensions, product improvements, new products for the current market, new products for a new market (Heany, 1983)
- Incremental newness of the technology embodied in the product: minor improvement, major enhancement, new related technology, and new unrelated technology (Meyer and Roberts, 1988)
- Original new products, reformulated new products (Yoon and Lilien, 1985)
- Generic product development map: development work, engineering prototype, core product, enhanced product, customized product, cost-reduced product, hybrid product (Wheelwright and Sasser, 1989)

Underlying the selection of the 12 products in this study were four families of products. Within each family, the products included the "first of family" in order to capture the learning from product to product. Newness was the degree of similarity of a product to other members of its family. Measurement of newness included the degree to which preexisting product parts, process equipment, tooling and manufacturing methods were altered or redesigned to suit the requirements of the product under development. The basis for this measurement system was a set of ten major electronic and mechanical/plastic product elements or parts common to all terminals. Rating scores for newness were based on a 5–point scale, with lower scores given for newer elements and higher scores for less-new elements. This measurement system provided a single newness score for each product, through addition of ratings for each product element.

MANUFACTURING ENGINEERING ACTIVITIES IN THE NEW PRODUCT DEVELOPMENT PROCESS

The company in the study saw the product development process as a set of five distinct stages: initiation, definition, development, verification, and the first year of routine volume production. At the end of each stage, a formal review of progress was held and "prime responsibility" for achievement of the targets for the next stage was transferred from one functional group to another.

ME first became involved in the product development process after the initiation stage, when the product concept was approved, and the focus had changed to definition of technical specifications. ME remained involved in the process through the development of prototype units, until the design conformed to product specifications and product cost targets were achieved in a pilot run at the end of the verification stage. Subsequently, a group of ME staff, separate from those involved in the development of the product, took responsibility for the support of the product in volume production. The prime interest in this study was on the development rather than the support group.

The development focus of ME changed over the course of the development process. The definition stage required the delivery of the following items by ME:

- Manufacturing/test plan
- Value analysis results
- Yield targets
- Subsystem cost targets
- Capital equipment and capacity requirements

During the development stage, ME carried out activities to complete the following deliverables:

- New process qualification
- Product costs
- Manufacturability assessment
- Manufacturing start-up program
- Inventory requirements (short term)
- Yield predictions

During the verification stage, ME was responsible for achievement of the following deliverables:

- Manufacturing yield update
- Product costs update
- Manufacturability update

Finally, after the start of routine volume production, the following deliverables were required from the ME support group:

- Manufacturing yields
- Product costs

Examination of the activities carried out and man-hours expended by ME in various stages of the development process indicated that over 50% of all ME activity was carried out during the development stage. However, ME began nearly two-thirds of these activities during the definition stage, while completing these activities in the verification stage. These findings are summarized in Tables 10.1 and 10.2, adding substance to the notion of overlapping phases of development described by Clark and Fujimoto (1989).

Table 10.1. Manufacturing Engineering Activity Content of Development Stages

Class of ME activity	Development stages			
	Definition	Development	Verification	Production
Number of activities started	24 (60%)	9 (22.5%)	4 (10%)	3 (7.5%)
Number of activities in process	24 (60%)	25 (62.5%)	21 (52.5%)	6 (15%)
Number of activities completed	8 (20%)	8 (20%)	18 (45%)	6 (15%)

MANAGING ENGINEERING CHANGE

For each division, managing engineering changes to product specifications, process equipment, tooling and methods was a major challenge. Cost savings from individual changes were often less than implementation costs, especially after the start of routine volume production. Further, the negative cost implications were greater for manufacturing operations characterized by higher unit volumes. The sheer number of changes often led to materials management problems. The president had become involved in a critical evaluation of the change process and the whole rationale behind the incidence of engineering change. Correspondingly, the company was changing its approach and attempting to limit the scale and scope of changes permitted after start-up.

Table 10.2. Phasing of Manufacturing Engineering Deployment on Nine Projects[a]

	Project place in product family	Percentage of ME manhours per stage		
		Definition	Development	Verification
Division				
A	1	25.90	37.33	36.76
	2	23.04	43.66	33.31
	3	27.80	38.82	23.15
	4	5.88	60.67	33.45
B	1	8.1	66.05	25.85
	2	8.1	66.05	25.85
	3	23.49	51.75	24.76
C	1	0.48	61.34	38.18
	3	6.46	60.99	32.55
Average man-hours per project		14.36	54.07	30.43
Standard deviation		10.48	11.50	5.38

[a]ME man-hour expenditures over the duration of each project were available only for nine of the products developed in three divisions.

Table 10.3. Company Standard Classification Scheme for Engineering Change

Class	Definition
1	An inoperative or potentially hazardous condition
1A	An inoperative or potentially hazardous condition in certain applications only
1B	An unsatisfactory condition that may be allowed to exist on a temporary basis
2	An improvement in design, but, in so doing, the design intent is affected
3	To introduce new features or to change the product rating
4A	Changes that do not affect design intent such as component substitutions, artwork recycles, or cost reductions
4B	Changes that do not affect design intent that must be applied as soon as possible
4C	Changes that do not affect design intent but improve marginal design conditions

In general, engineering change after the start of routine volume production had three characteristics: motivation, anticipation, and transparency.

Motivation for Change

Manufacturability-related change was motivated by the need for materials substitution, cost reduction, design corrections, design improvement, documentation change, new features or yield improvement. To control such complexity, the company operated a change classification scheme that grouped all engineering changes according to specific criteria. This classification scheme is summarized in Table 10.3. Manufacturability-related changes were identified as class 4 changes. While all divisions used the classification scheme, not all of the change activity within the divisions was captured by the scheme. Discussion with managers helped to identify additional instances of manufacturability-related change. For example, Division A introduced extensive process automation for cost reduction purposes, after the start of volume production of product A1. The automation was carried out as a series of cost reduction projects, but did not register in any class 4 change notices.

Change Anticipation

Manufacturability-related change within the first year of volume production was either planned or unplanned. Unplanned engineering change occurred in many areas, as shown in Table 10.4, but one of the most common areas was materials substitution. The ME managers traced the need for materials substitutions back to the validity of manufacturability assessments made during the development process. The batch size of components or materials provided to ME and manufacturing for assessment of manufacturability may not have been large enough to capture the range of variations in specification with impact on manufacturability. Only in volume production did such variations emerge, and with them the requirement to make unplanned materials substitutions.

Unplanned change in methods occurred for seven products; in three cases methods were changed for cost reduction or yield improvement. Methods change was largely unplanned before the verification stage, although ME "product support" was

Table 10.4. Products for Which Manufacturability-Related Engineering Changes Were Made During the First Year of Volume Production

Change area	Number of products changed	Motivation for change						
		CR[a]	MS[b]	DC[c]	DI[d]	DO[e]	NF[f]	YI[g]
Unplanned changes								
Product specification	11	6	8	7	8	2		
Process equipment	2	2						2
Tooling	1	1						
Methods	7	3		2				3
Planned changes								
Product specification	2	1	1					2
Process equipment	6	6						2
Methods	2	1						
Class 4 changes only								
Product specification	11	6	7	5	3	2		
Methods		1						

[a]CR = cost reduction.
[b]MS = material substitution.
[c]DC = design correction.
[d]DI = design improvement.
[e]DO = documentation change.
[f]NF = new feature.
[g]YI = yield improvement.

a planned activity that included handling unplanned demands for methods change. Thus, while the specific methods changes were unplanned, and ME was reactive in making these changes, the provision of resources for shop support and cost improvement was proactive in anticipation of the need for change.

Planned change occurred in at least six products, mainly in the area of ongoing process automation for cost reduction, as summarized in Table 10.4. Simultaneous development of a product and its associated processes was fundamental to the development of manufacturable new products. Process automation was carried out as a series of cost reduction projects. This automation was planned during the product development process, and typically, development had commenced before the start of volume production. However, implementation and start-up of the automation was consciously deferred until the manufacture of the product stabilized on the existing line, or on a largely manual line. The possibility of "line-stopping" start-up problems was felt to increase with the newness of the product and process technologies. In some cases, a staggered start-up of a new product and its associated new process was also the result of a resource shortfall within the ME organization. Through an inadequate staffing policy, the ME staff available were mismatched with the scope and scale of the development activities. As a result, the product was designed for both manual and automated assembly, and the initial costs were higher than they would have been otherwise.

Separating class 4 changes from the above planned and unplanned changes indicates that class 4 changes were recorded for 11 of the 12 new products studied. These changes were made during the first year of routine volume production, that is, after the verification stage. (No data were available for one product as volume production had commenced during the month preceding this study.) The separation reveals that most class 4 changes were unplanned, related to product specification, and were motivated primarily by materials substitution, cost reduction, and design correction. The spread of these manufacturability-related changes is summarized in Table 10.4.

Change Transparency

Each product development project involved design and development of both product and process specifications. In each area, but most especially the product specification, engineering change had the potential to change the design intent of the product. The design intent described the range of operating features and compatibilities that made the product attractive to buy, and easy to install, use, service and upgrade. Change that affected these features and compatibilities was apparent to the ultimate product user, the customer, and to the intermediate user of the product specification, manufacturing. As indicated in the company change classification scheme, class 1, 2, and 3 changes affected design intent. Among these classes, a class 1 change, made in response to an inoperative or potentially hazardous condition, was the most serious: if the product feature did not work, the design intent was not being fulfilled. If this change was required after the product had been released to the market, costs were incurred in image and reputation, in addition to more quantifiable terms as redesign costs and lost revenues.

Change affecting the design intent was not always transparent to manufacturing, especially if the change was to materials, process equipment, tooling, or methods. Implementation of such change resulted in additional manufacturing costs measured in terms of waste, rework, and downtime arising from stopping and scrapping work in process, both unlearning and relearning of methods, and qualification of new parts or suppliers. The company estimated that the relative cost impact of a design flaw detected after the verification stage was 100 times greater than one detected in the development phase.

Change that did not affect the design intent of a product was transparent to the customer, but not necessarily to manufacturing. This type of change affected manufacturability through re-layout of printed circuit boards to reduce component density, substitution of components, automation of process stages, or even reversion to manual methods. When carried out reactively, the objective of this type of change was to meet cost or yield targets originally set as part of the original product development objectives. When carried out proactively, these changes were expected to improve cost or yield performance relative to targets already attained.

ME DEPLOYMENT AND THE INCIDENCE OF ENGINEERING CHANGE

The low number of products in the study allowed the development history of each product to be investigated in depth. In addition, where data were available, context-related differences in, and associations between, the incidence of class 4, or manufacturability-related engineering changes, and ME deployment decisions were analyzed further using statistical techniques appropriate for small samples. These techniques are described by Siegel and Castellan (1988), and included the Mann-Whitney U-test of difference, and the calculation of the Kendall rank-order coefficient of association, tau.

The development context of the product development projects was defined by the degree of product and process newness. Products were grouped as newer and less-new relative to the median value of the measure of product newness. Typically, the "first of family" was newest, while subsequent models varied in their relative newness. In contrast to less-new products, newer products included more new components and assemblies that had not been assessed for manufacturability previously and had not been tried in volume production. Sorting out the resulting uncertainties in product and process specifications extended throughout the development process, which changed the development task for ME.

ME Staff Experience and Manufacturability-Related Change

In each division, ME staff gained experience over a number of product development projects. This experience translated not just into the ability to make competent choices when matching product and process specifications, but also into an understanding of the development process. On the basis of this understanding, the ME staff provided positive guidance in developing manufacturable product designs in the context of existing manufacturing processes. However, this understanding was easily lost, as it was held by ME staff who could leave the product development project "without trace of their being."

In general, reduced incidence of engineering change was associated with the assignment of more experienced ME staff to a project (tau = -.5511, p < .015) for the 11 projects for which these data were available. These staff brought a depth of understanding of the product and process technologies to bear on the design tasks. The learning from product to product was captured in the use of common parts modules, formal revisions to parts drawings, component specifications, manufacturability guidelines, and phasing of assessments. This learning was also absorbed by individual ME staff.

The importance of ME staff experience in relation to the avoidance of engineering change has strong implications for the staffing of the ME function. The vulnerability of ME staff experience was illustrated in three of the divisions. While there were no explicit statements of manufacturing strategy, each division had de facto policies that reflected the particular product markets served and their competitive realities. However, in two divisions, operational pressures strongly influenced the policy on ME staff size and composition. Short-term responses to operational pressures led to "swings" affecting the ME establishment, both in size and mix of engineers and technicians, sometimes resulting in inconsistency between the size and composition of the ME staff and the divisions' long-run objectives.

When viewed in the context of the association between the incidence of manufacturability-related engineering change in newly developed products and ME staff experience, these swings in the ME profile had potentially more damaging consequences. Rather than just reducing numbers of ME staff and limiting the scope or sequencing of projects undertaken, these swings helped to dilute the very experience required to avoid costly engineering change. For example, while ME staff were involved in each project team, junior manufacturing engineers were assigned to some less-new products. In doing this, the divisions failed to capitalize on the experience gained by senior ME staff on earlier projects. Thus, ME staffing decisions seem to have long-term and quantifiable implications for the performance of product developments.

Emphasis on Manufacturability and Manufacturability-Related Change

By definition, newness implied a diminished use or availability of existing experience both built into product and process elements and held by the ME staff allocated to the development project. Newer products included more new components and assemblies that had not previously been assessed for manufacturability.

Statistical analysis of the differences between newer and less-new products suggested that, for newer products, ME placed greater emphasis on manufacturability ($U = 5$, $p < .02$) and placed that emphasis earlier in the development process ($U = 6$, $p < .03$). However, there was evidence to suggest that the incidence of manufacturability-related change was higher for newer products ($U = 10$, $p < .18$).

The ME staff placed more emphasis on manufacturability earlier through developing and applying design rules and through carrying out manufacturability assessments during the product development process. However, because of the newness, these components and assemblies had not been proven in volume production, as those in less-new products would have been, and so required engineering change to effect necessary improvements.

Yet, in spite of the limitations of the assessment process, the performance of newer products seemed amenable to the emphasis placed by ME on manufacturability. Among the products for which data were available for statistical analysis, six products were classified as newer. For these newer products, the incidence of engineering change was lower for greater emphasis on manufacturability (tau $= -.4140$, $p < .126$). In contrast, for less-new products, there was no evidence of association between the incidence of engineering change and ME emphasis on manufacturability. As they included many proven components and assemblies, these products seemed to require less emphasis on manufacturability and showed no response to such emphasis.

Phasing of ME Deployment and Manufacturability-Related Change

The differences between newer and less-new products were maintained on investigation of the expenditure of ME man-hours as an indicator of the phasing of ME deployment in the product development process.

For newer products, there was no evidence of association between the incidence of engineering change and expenditure of ME man-hours in the definition stage. In contrast, the incidence of engineering change was higher when ME man-hour expenditure was higher in the definition stage for the four products classified as less-new for which these data were available (tau $= .6667$, $p < .87$).

This difference was explained in terms of the use of the design information necessary for ME to carry out its product development responsibilities. This infor-

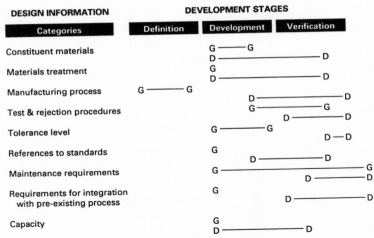

Figure 10.2. Category and completeness of information available to manufacturing engineering for products developed first in their families. G = general information available; D = detailed information available.

mation evolved from "general" to "detailed" over the course of the product development process, as shown in Figure 10.2. While some detailed design information was available as early as the definition stage for those products that were first in their respective families, a complete set of detailed information was not available in all categories until the verification stage. As such, newer designs were not "frozen" until late in the development process, and there was no particular relationship between engineering change and early expenditure of man-hours.

In contrast, as less-new products were variations on earlier models, design details were carried forward as candidates for re-use. Where ME viewed these products largely as repeats, they expended a larger proportion of man-hours in the definition stage, during which they froze many of the design details carried forward. As a result, these details were found to have been frozen on the basis of incomplete understanding of their implications for the revised product. The design limitations became apparent after the start of volume manufacturing and resulted in engineering changes.

For example, details of the plastic cover and base of a terminal were often finalized and frozen before the circuit designs were completed because mold design and proving were lengthy, costly activities. Subsequently, circuit designers faced fixed space limitations when locating circuit boards within the cover. However, these spatial constraints and the high costs of changing molds forced compromises on board manufacturability that became apparent only after volume production started and precipitated engineering changes.

A lower incidence of change was associated with higher expenditure of ME man-hours in the verification stage for the eight projects for which these data were available (tau = − .4444, p < .066). The strength and significance of this basic relationship was moderated to some extent by product newness. However, the direction of the association was unchanged. Such later ME involvement resulted, for example, in more complete process documentation, detailing standards and methods for the shop floor. The completeness of this documentation avoided issuing engineering changes later.

IMPLICATIONS FOR MANAGEMENT OF PRODUCT DEVELOPMENT

The Earlier the Better?

Managing the product development process for manufacturability is not a simple task. There are no general rules, especially at the level of detail at which ME operates. The findings of this study suggest that a general recommendation to involve ME earlier in the development process is unhelpful; the relationship between ME deployment and product performance is more complex than that and reflects the relative newness of the product.

There is no question that ME has a role in the development of manufacturable new products. However, in playing that role, the ME manager must have a keen awareness of the development context of the product. Product newness has an impact on both the performance of newly developed products and on the deployment of ME staff during the development process.

First, expending a high proportion of ME man-hours too early in the development process, in the definition stage, seems to lead to more manufacturability-related changes for less-new products. The MEs may freeze the design prematurely on the basis of incomplete and unstable information on the interrelationships among components and subassemblies. However, manufacturability-related changes seem fewer when MEs expend a high proportion of man-hours later, during the verification stage.

Second, ME appears to place more emphasis on manufacturability and to do so earlier for newer products. Although more emphasis tends to decrease manufacturability-related changes for newer products, generally speaking, the overall incidence of manufacturability-related changes is higher for these products.

In light of this, the ME manager should consider carrying out an evaluation of product newness at the start of the definition stage of the development process. This evaluation would parallel manufacturability assessment and value analysis sessions and would establish the newness of the product and its processes, relative to its family, in terms of major product components and subassemblies. The focus would be on the management implications of newness for ME deployment. This family, rather than product focus, would also put the performance of the resulting product in proper perspective: that is, in relation to its family.

Right-First-Time or Right-Next-Time?

Design right-first-time is an unambiguous objective; however, its achievement is difficult. The observed incidence of engineering changes seemed to run counter to the company objective of design right-first-time in spite of management exhortations, procedures, and earlier manufacturability assessments. The question remains, then:

Is an objective of design right-first-time attainable through deployment of ME staff, or is a more practical objective to design right-next-time?

Engineering change may occur at any time before or after the start of volume production. Before the start of production, change is an expected intermediate outcome of the product development process, as the imprecision, risk, and uncertainty are extracted from the design specification by the project team through design reviews, manufacturability assessments, and interfunctional cooperation. Ideally, the incidence of change should have been reduced to zero by the start of volume production. However, the experience of the four divisions in this study was that, while some change affecting the design intent of products occurred after the start of volume production, the majority of the change did not affect the design intent. As such, from a design intent perspective, the products were designed right-first-time.

In contrast, the incidence of manufacturability-related change introduced after the start of volume production suggested that the products were not designed right first time. The time and volume-based limitations on the screening of designs to eliminate these changes during the development process suggests an inevitability of this type of change after the start of volume production. As such, change that does not affect the design intent seems almost a natural outcome of the move to volume production.

However, consideration of the role of manufacturing engineering and the impact of the development context suggests that such change is manageable and can be reduced in part, if not to zero. Underlying the selection of the 12 products in this study were four families of products. The newer products were less manufacturable and experienced more manufacturability-related change than the less-new products. Further, the areas where less-new products experienced problems were not features of earlier products in the same family. Therefore, through addressing the manufacturability-related change in the earlier members of these product families, the recurrence of that type of change in later, less-new products was avoided. In effect, the divisions learned from their previous change experiences and carried that learning over to later products.

Accordingly, for product features such as manufacturability, apparent only to manufacturing, a more appropriate objective seems to be to design right-next-time. By addressing and learning from those manufacturability problems that emerge in newer products after volume production starts, their recurrence in later, less-new products can be reduced.

IMPLICATIONS FOR MANUFACTURING ENGINEERING MANAGEMENT

The study indicates that using experienced MEs tends to decrease manufacturability-related changes, but operational pressures lead to unsystematic assignment of MEs, so that there is no follow-through within the same product family. These findings carry several implications for senior manufacturing managers, and for manufacturing engineering managers. For each manager, these implications involve manufacturing policy, organization structure, and product performance.

Implications for the Senior Manufacturing Manager

Policy formulation. As a policy maker for the manufacturing organization, the senior manufacturing manager plays a key role in the policy formulation process and oversees its implementation. This planning process aims to support a variety of strategic targets, such as market expansion, increasing market share, improving margins, and building the manufacturing team and improving its skills. The context within which new products are developed is a related strategic issue and includes the newness of the product and process technology. Newness shapes the management task and the performance outcome and, so, is important to the planning agenda.

Without close interaction with the ME manager during the policy formulation process, the senior manufacturing manager risks losing the benefit of accumulated ME experience and may reach decisions about the development context incompatible with the selected business targets. ME managers are aware of the issues in developing manufacturable new products. They understand these issues, are exposed to them every day, and live with their consequences.

Organization structure. ME has a key "line" rather than "staff" role in the product development process. For the senior manufacturing manager, this role raises questions of where ME fits in the manufacturing organization and what staffing policy is appropriate for building, maintaining, and using ME's accumulated experience. Such a reevaluation may lead, in the extreme, to the integration of production and engineering activities such that engineering managers have line responsibilities, with production managers providing staff support.

Yet while the importance of ME to product development is increasing, ME's organizational identity is becoming harder to define. Earlier involvement of ME in the product development process blurs the distinctions between functional, discipline-based design and manufacturing engineers. Effective product development requires ME and design engineers to be familiar with each other's responsibilities, competencies and limitations. While manufacturability guidelines reflecting manufacturing process capabilities build this familiarity, they are not sufficient. ME needs to become familiar with the constraints on product design, so that the guidelines do not place unnecessary limitations on the product designers. Accordingly, senior management should carry out routine assessments of formal training needs for manufacturing engineers. Similarly, the design engineers need to appreciate the processes used to manufacture their designs. An ongoing series of structured visits to the facilities where their products are manufactured is one way to aid this appreciation.

Traditionally, senior management has viewed ME as an element of manufacturing overhead, whose size is determined through financial cost ratios rather than product and process responsibilities. However, short-term reductions in ME staff in response to current sales results lead to inconsistency between the size of the ME staff and the firm's long-run objectives. Ensuring that both ME staff level and mix are consistent with the development program of the firm is important to the senior manufacturing manager. Without the appropriate numbers, mix of expertise, and experience, ME will be forced to "cut corners" in trying to develop products and processes to specification in the increasingly shorter times available. The result will be costly manufacturing problems, engineering changes, and loss of intellectual assets. Not all

ME experience is captured in documentation or training programs and so is vulnerable. Maintaining the continuity of ME membership on successive product development teams is one way to capitalize on past experience and learning and reduce vulnerability.

Performance. As a motivator, the objective of design right-first-time is unambiguous. However, as noted earlier, its achievement is difficult. Accordingly, the length of time over which manufacturability-related product performance is evaluated and the basis for that evaluation are worthy of senior management attention. Newer products are inherently less manufacturable and experience more manufacturability-related change than less-new products. Ideally, the incidence of this type of change should have been eliminated by the start of volume production. In practice, however, manufacturability-related change is common after production starts and is often related to time and volume-based limitations on the screening of designs during the development process.

Each product exists in relation to its family and is a function of these related products. By addressing and learning from those manufacturability issues that emerge after volume production starts, the recurrence of that type of change in later, less-new products might be avoided. Accordingly, evaluation of the manufacturability of a product in relation to its family, and over a longer period, might provide clearer guidance for future action.

Implications for the Manufacturing Engineering Manager

Policy implementation. The ME manager plays a key role in the implementation of manufacturing policies. Managing the product development process to achieve product manufacturability and avoid engineering change is not simple, especially given the impact of the development context on the performance of new products and on the deployment of ME staff. The implication is clear: assessment of the development context of a project is important before committing any ME resources to the project.

Organization of ME. Earlier involvement of ME in the product development process requires that both ME and design engineering become familiar with each other's responsibilities. A disciplined management approach to the coordination of design and ME resources during the product development process contributes to achieving this familiarity. However, such an approach does not, of itself, replace the need for individual ME staff members to be able to relate to other functionaries without guidance from procedures. Accordingly, the ME manager should consider the development and assignment of ME staff, not merely in a technical engineering sense, but also in relation to their ability to interact and to integrate their activities with product designers and marketeers. Development of these ME staff will be helped by use of explicit criteria for manufacturability and care in the phasing of their use during the development process.

The need for this type of integration and reflection defines a new breed of ME. No longer just the technical specialist in specific process areas, the ME should have strong communication skills, work effectively in multidisciplinary teams, be able to communicate management experience, and feel challenged to update the breadth and depth of both technical and managerial skills on a regular basis.

Performance. Providing a forum for ME staff members to structure and disseminate their experience following each product development project is an important consideration for the ME manager. Structured ME experience of the product development process is captured most vulnerably by ME staff members. The vulnerability arises from the difficulty in achieving continuity of ME staff membership of successive product development teams, owing either to competing projects or to changes in number and mix of engineers and technicians resulting from short-term changes in sales levels.

Part of the structuring of experience could come from analysis of the records of ME man-hours expended during the development process and from engineering change notices. ME managers have details of man-hour expenditure available to them, gathered as part of the cost control function and based on time-sheets. Engineering changes are recorded for change control purposes. These data have uses, beyond change or cost control, such as project evaluation and improvement. While managers may say, with some truth, that these data do not reflect actual occurrences, because they are used to smooth budgets or to control the extent to which change is publicized, they are only fooling themselves. Inattention to the quality of man-hour expenditure and change records results in a poor basis from which to improve performance on future projects.

IMPLICATIONS FOR RESEARCH

A host of tools is available, and under development, to improve the responsiveness, flexibility, and effectiveness of manufacturing in relation to product development. Many of these tools are computer-based and integrated with systems in other functional areas. However, without a fundamental understanding of the management of ME resources, firms may never realize the potential of these tools for improvement. The challenge facing researchers is to facilitate such understanding.

This study focused on the deployment of ME resources during an in-house development process. The role of an ME group with an external focus on vendor management was not investigated. The parallels between the management of this externally focused group and the ME group investigated in this study should be explored. What are the similarities and differences in the management task of integrating design with manufacturing, when the manufacturer and ME are geographically and organizationally separated? Are the same engineering resources of use, and are the deployment strategies similar? As companies change the degree to which they depend upon outside sources for the supply of components and subassemblies, these questions become key in making the best choices among alternative types of engineering resources:

> The premier manufacturers in the world—IBM, GE, Toyota—send their own manufacturing engineers out to their suppliers to help them improve their processes, improve designs, and improve quality, reduce costs and improve schedule compliance. If a firm

has too few manufacturing engineers in its own plants to do that for itself, how is it going to do that for 20 or 200 of its suppliers? (Gunn, 1987, 106–107)

ME is a strategically important function in the fight by companies to develop a sustainable, manufacturing-based competitive advantage. The focus of this chapter was on the product development activities of ME. There are other dimensions to this function, including process development, support, and improvement and product support and improvement. Exploration of these dimensions of ME management is a necessary further step towards understanding a major infrastructural element of manufacturing.

ACKNOWLEDGMENT

This research was supported, in part, by the National Centre for Management Research and Development, Canada.

REFERENCES

Andrew, C. G. "Engineering Changes to the Product Structure: Opportunity for MRP Users." *Production & Inventory Management*, 3rd Qtr., 1975, 76–86.

Clark, K., and T. Fujimoto. "Overlapping Problem Solving in Product Development." In *Managing International Manufacturing*, ed. K. Ferdows. Amsterdam: North Holland, 1989, 127.

Cooper, R. G. "The Components of Risk in New Product Development: Project New Product." *R & D Management* 2(2), 1981, 47–54.

Coughlan, P. D. *Proceedings of ASAC 1989 Conference* 10(7), 1989, 10.

Daetz, D. "The Effect of Product Design on Product Quality and Product Cost." *Quality Progress*, June 1987, 63–67.

Dewhurst, P., and G. Boothroyd. "Design for Assembly: Automatic Assembly." *Machine Design*, January 26, 1984, 87–92.

Diprima, M. "Engineering Change Control and Implementation Considerations." *Production & Inventory Management*, 1st Qtr., 1982, 81–87.

Duck, T. "Design for Manufacturing Integration." *Production Engineer*, September 1986, 48–51.

Gerstenfeld, A. "A Study of Successful Projects, Unsuccessful Projects, and Projects in Process in West Germany." *IEEE Transactions on Engineering Management* EM-23 (3), 1976, 116–123.

Gunn, T. C. *Manufacturing for Competitive Advantage.* Cambridge, Mass.: Ballinger, 1987.

Hales, H. L. "Producibility and Integration: A Winning Combination." *CIM Technology*, August 1987, 14–18.

Hauser, J. R., and D. Clausing. "The House of Quality." *Harvard Business Review* 66(3), 1988, 63–73.

Hayes, R. H., and K. B. Clark. "Exploring the Sources of Productivity Differences at the Factory Level." In *The Uneasy Alliance: Managing the Productivity-Technology Dilemma*, eds. K. B. Clark, R. H. Hayes, and C. Lorenz. Boston: Harvard Business School Press, 1985, 151–158.

————, S. C. Wheelwright, and K. B. Clark. *Dynamic Manufacturing: Creating the Learning Organization.* New York: Free Press, 1988.

Heany, D, F. "Degrees of Product Innovation." *The Journal of Business Strategy*, Spring 1983, 3–14.

Imai, K., I. Nonaka, and H. Takeuchi. "Managing the New Product Development Process: How Japanese Factories Learn and Unlearn." In *The Uneasy Alliance: Managing the Productivity-Technology Dilemma*, eds. K. B. Clark, R. H. Hayes, and C. Lorenz. Boston: Harvard Business School Press, 1985, 337–375.

Langowitz, N. S. "An Exploration of Production Problems in the Initial Commercial Manufacture of Products." *Research Policy* 17, 1988, 43–54.

————. "Managing New Product Design and Factory Fit." *Business Horizons* 32(3), May-June 1989, 76–79.

Meyer, M. M., and E. B. Roberts. "Focusing Product Technology for Corporate Growth." *Sloan Management Review*, Summer 1988, 7–16.

Noaker, P. M. "CIM/Software: This Software Accelerates Information Flow." *Production*, June 1987, 62–64.

Rohan, T. M. "Designer/Builder Teamwork Pays Off." *Industry Week*, August 7, 1989, 45–46.

Siegel, S., and N. J. Castellan, Jr. *Nonparametric Statistics for the Behavioral Sciences*, 2nd ed. New York: McGraw-Hill, 1988.

Starr, M. K. "Modular Production—A New Concept." *Harvard Business Review* 43(6), 1965, 131–142.

Szakonyi, R. "Mechanisms for Improving the Effectivenss of R & D: How Many Mechanisms Are Enough?" *R & D Management* 15(3), 1985, 219–225.

Takeuchi, H., and I. Nonaka. "The New Product Development Game." *Harvard Business Review* 64(1), 1986, 137–146.

Walleigh, R. "Product Design for Low-Cost Manufacturing." *The Journal of Business Strategy,* July-August 1989, 37–41.

Wheelwright, S. C., and W. E. Sasser. "The New Product Development Map." *Harvard Business Review* 67(3), 1989, 112–125.

Whitney, D. E. "Manufacturing by Design." *Harvard Business Review* 66(4), 1988, 83–91.

Wood, A. R., and P. D. Coughlan. "Manufacturability and New Product Introduction Procedures." *Proceedings of ASAC 1988 Conference* 9(7), 1988, 91–100.

Yoon, E., and G. L. Lilien. "New Industrial Product Performance: The Effects of Market Characteristics and Strategy." *Journal of Product Innovation Management* 3, 1985, 134–144.

11
MANUFACTURING FOR DESIGN: BEYOND THE PRODUCTION/R&D DICHOTOMY

KIM B. CLARK, W. BRUCE CHEW and TAKAHIRO FUJIMOTO

Intense international competition in the 1980s has focused increasing attention in both the academic and popular press on fundamental sources of superior performance in manufacturing. Sizable gaps in productivity and quality between U.S. manufacturers and their Japanese and European competitors in a wide range of industries led to new perspectives on the role of inventory in production, the impact of statistics on process control, and the central role of committed, capable people in solving problems and achieving continuous improvement.

But the search for an understanding of manufacturing performance also led outside factory walls to offices and laboratories in which decisions about product design are made. There is now abundant evidence that the design of a product, including its basic architecture, decomposition into subsystems, and choice of basis components, as well as detailed engineering decisions about the shape and number of parts and method of fastening have a critical impact on manufacturing cost and quality. This understanding has begun to have real effects. Recognition of the impact of design on manufacturing has not been limited to the academic literature and the lecture circuit. Methods and techniques for developing designs that are easier to assemble and use far fewer indirect resources have been implemented with well-documented effects.

Along with many other changes in manufacturing management in the 1980s, a renewed emphasis on design and manufacturing began to narrow competitive gaps in productivity and quality. But the 1980s also witnessed the emergence of other dimensions of competition, particularly the timely, effective development of new products, that aggressive competitors have used for advantage in a variety of industries. As competitive gaps narrowed in production, the performance of design and engineering has become a focal point of competitive rivalry. Indeed, in industries such as automobiles, home appliances, medical products, engineering workstations, and copiers, rapid and efficient development of attractive new products has become a competitive imperative.

Our research on manufacturing and technology strategy over the last several years has followed a similar path, and, in a sense, has come full circle. We began the decade of the 1980s concerned about factory level productivity and quality. Our comparative studies of factories in the United States, Japan, and Europe uncovered striking differences in the architecture of manufacturing and in basic principles of manufacturing management. But it was also clear that design and development of new products played an important role in quality and productivity. We therefore launched a study of the development process, and soon broadened our focus to include the competitive role of development and its own performance in terms of lead time, engineering productivity, and design quality (Clark, Chew, and Fujimoto, 1987). It was here, in our search for an understanding of superior performance in development, that we came full circle: the impact of design on manufacturing was our starting point (and continues to be a critical connection), but we have found that manufacturing capability has an important impact on the design process and the performance of product development. Critical manufacturing activities in development—i.e., making prototypes, building tools and dies, pilot production, and manufacturing ramp-up—can have a significant impact on lead time, cost, and total product quality.

We examine this impact here using data developed in our study of the world auto industry. We first lay out a framework for linking manufacturing capability to development performance. The central notion is that the heart of development is a sequence of design-build-test cycles. Excellent performance in building and testing can have an important effect on the speed and quality of problem-solving. We then review overall development performance in our data, including comparative analysis of lead time, engineering productivity, and total product quality.

Next, we examine different approaches to building prototypes and tools and dies. These activities are on the critical path of vehicle development programs (in terms of both quality and time), and performing them faster and better has an impact on overall development performance. Then we look at manufacturing activities at the tail end of development: pilot production and ramp-up. Because investments are high at this stage and because early customer experience can have an important effect on reputation and long-term market success, completing these activities quickly and well is crucial to the success of a development program. The chapter concludes with observations about the traditional dichotomy between management of production and management of R&D and suggests important common themes that characterize outstanding performance in both domains.

THE FRAMEWORK

The design and development of a product is an iterative process of successive refinement. The process typically begins with concept definition, in which planners and engineers lay down the architecture of the product and establish broad objectives such as target market, price range, performance parameters, customers, and relationships to other products. Over a period of weeks and months the concept is successively defined and refined through product and process engineering. In these stages of development, early models and concepts are developed through drawings, clay models, full-size mock-ups, and prototypes. Once established, the basic design moves through pilot production, ramp-up of volume production, commercial sale, and customer use. As this process unfolds the embodiment of the design changes. It goes from an informational representation (whether a drawing or electronic code is unimportant here) to physical representation (a prototype, for example) and back to informational representation for further refinement. Eventually the design is embodied not only in product blueprints or a CAD database, but also in tools and dies and procedures and work practices on the shop floor.

At each stage of the development process, engineers and designers confront a set of objectives and alternate ways of meeting them. Although some alternatives are well defined and some choices obvious, many objectives cannot be achieved with obvious alternatives. Framed in these terms, a gap between objectives and the performance of obvious alternatives defines a design problem. When this happens, designers and engineers launch what we call a problem-solving cycle. Each cycle involves framing the problem, generating alternative solutions, building a model, running experiments, and evaluating the results. The process is cyclic because typically the results of tests and experiments generate new understanding that leads to different alternatives (if not a reframing of the problem) and further models and tests.

Problem-solving thus rests on three critical activities: (1) design—framing the problem and devising alternative solutions; (2) build—making models of the part or the product (either in hardware or in software); and (3) test—setting up experiments, calibrating, measuring, and evaluating. These design-build-test cycles occur on a small scale (an individual engineer, a small group) and on a large scale (major component departments). In product engineering, for example, engineers take target performance specifications, styling models, and product layout information from the product planning stage as input or objectives and generate alternatives in the form of drawings. Prototypes are built according to the drawings and tested in the laboratory, on the test track, and on the road. Underlying this overall process are literally hundreds of problem-solving cycles connected to specific parts and components.

The performance of the design and development process depends on the way the development organization executes and integrates numerous design-build-test cycles. Although several aspects of that execution and integration are critical, we focus here on the impact of manufacturing capability on the build dimension of problem solving. We focus in particular on the effect of superior build performance on three dimensions of performance.

Lead Time

We define overall lead time as the time from the beginning of concept generation to market introduction. Other lead times—e.g., for engineering—may be important in assessing development performance. Lead time affects the competitiveness of a product at introduction because it determines how far in advance of market introduction critical design decisions must be made. In a rapidly changing market, long lead times can be a serious disadvantage. Lead time also affects how much time engineers and designers have to refine the product. Thus, lead time that is too short (given engineering capabilities) may result in a product that is not well developed.

Total Product Quality

We look at two dimensions of quality. The first is conformance—how well the actual product conforms to the design once it has been manufactured. The second is the quality of the design itself—how well the design (when made with exact conformance) satisfies customer expectations. Together, these two measures define total product quality, the most important measure of the effectiveness of the development process.

Development Productivity

The overall cost of development has several components, but we focus here on engineering hours. This variable is measurable (and available in our sample), and engineers and designers are one of the most critical resources in development and often a crucial limit on total development output. Though development productivity has some effect on the total cost of the product, its real impact is on the number of projects that a given development organization can accomplish over a period of time.

The potential impact of superior build performance on lead time, total product quality, and development productivity is relatively straightforward. To the extent that building models faster reduces the time for executing a cycle, it may shorten the time to solve problems. If the particular problem in question is on the critical path of the project, the capability to build rapidly can affect overall lead time. But it may also affect the quality of the solution. A rapid build means that the feedback loop between design and test is shorter. A shorter feedback loop means that information will be fresher and the possibility of information decay reduced.

Quality, particularly conformance quality, also may be affected by the way models are built. If, for example, those who do the building catch processing problems with the design and have the capability to feed that information back to the engineers, they not only may save time by avoiding later engineering changes, but they may catch problems that otherwise would remain in the design through commercial production. In this sense, the quality of the build process may have an important influence on the ultimate quality of the parts, components, and total system.

A faster, better build capability, in addition to affecting lead time and quality, may affect the productivity of engineering as well. Part of the productivity effect arises from its impact on lead time and quality. Since a portion of engineering resources on a project is dedicated for the duration of the project, reducing lead time will directly reduce engineering hours. Further, a higher-quality build that helps to reduce engineering changes and downstream problems will result in fewer engineering hours and thus higher productivity. But a faster, better build capability also may have direct effects. For example, if models are built more rapidly with fewer problems, the engineers may learn more from each cycle, thus reducing the overall level of problem solving required to achieve project objectives. The effect would be to reduce engineering hours and improve productivity.

All of these effects seem logical, but whether they actually show up in practice is an empirical question. In order to see what effect differences in manufacturing capability might have, we turn to data on design-build-test cycles in our study of the world auto industry.

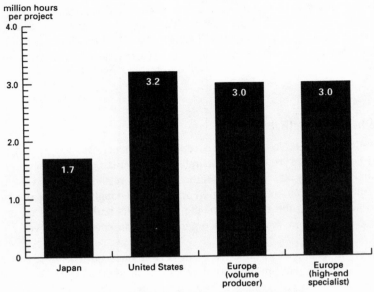

Figure 11.1. Engineering hours required to develop a $14,000 compact car with two body styles. Adjusted engineering hours are the hours required by the average firm in each region/group to design and engineer the average vehicle in the sample. Adjustments are made for the type of vehicle, use of pre-existing parts, and design hours spent by suppliers. Source: Based on regression analysis; see Clark et al. (1987) for details.

OVERALL DEVELOPMENT PERFORMANCE IN THE WORLD AUTO INDUSTRY[1]

Data in this section were developed as part of a larger study of product development in the world auto industry (see Clark and Fujimoto, 1991; Fujimoto, 1989). Our study of 29 automobile development projects around the world—12 in Japan, 6 in the United States, and 11 in Europe (for further information, see Clark, Chew, and Fujimoto, 1987)—found a wide range of performance in terms of cost, lead time, and quality. Data pertinent to cost are presented in Figure 11.1, which shows an estimate of the number of engineering hours required to produce an average vehicle's design. We present engineering hours rather than cost for several reasons. First, this labor productivity measure presents data without the interference of exchange rates and wage variation. Second, it is a significant cost in its own right: note that the units are millions of hours per project.

Average lead time is shown in aggregate in Figure 11.2 and by project in Figure 11.3. Finally, a variety of quality data is presented in Table 11.1 (note that the quality data pertain to the firm as a whole, and thus the number of observations in Table 11.1 (22) is different from the number of projects).

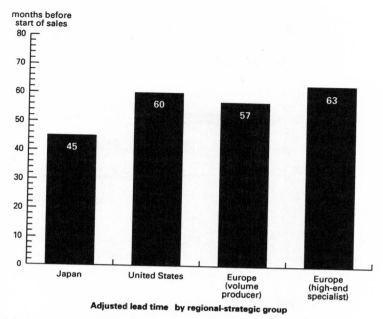

Figure 11.2. Lead time required to develop a $14,000 compact car with two body styles. Adjusted lead time is the time from concept development to market introduction required by the average company in each region/group to complete the average project in the sample. Source: Based on regression analysis; see Clark et al. (1987) for details.

These figures show a clear Japanese advantage in lead time, productivity, and conformance quality; the Japanese use over 40% fewer engineering hours on average to produce a comparable product. Lead time is also far better at Japanese firms on average; European and U.S. firms take a year longer to design a car. There are many possible explanations for this difference. The detailed data suggest that this advantage stems in part from the timing of different phases (more use of overlapping development) and in part from simply taking less time to accomplish critical tasks. The role of Japanese vendors is also significant (Clark, 1989). The quality data are more mixed; there is no overall Japanese advantage. There is a clear edge in conformance quality—measured by reliability and problems in the first few months of ownership. Further, a few Japanese firms have achieved high levels of design quality.

It appears, then, that Japanese producers have an advantage in executing and integrating design-build-test cycles. It is our belief that an important portion of the international differences in performance rests not on the design piece of that cycle but on build-test. That is, Japanese firms have been able to enhance their development performance by exploiting superior manufacturing capabilities.

Japanese firms have been noted for excellence in manufacturing for some time. But our discussion here is not limited to the often cited reduced inventory levels or superior defect rates in Japanese factories. When we speak of manufacturing capability we are referring to a broadly based set of skills with wide-ranging application. Our contention is that this broad-based manufacturing capability goes beyond simply

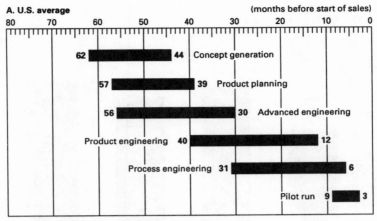

Note: average lead time of 6 U.S. projects

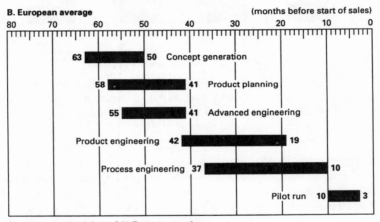

B. European average

Note: average lead time of 11 European projects

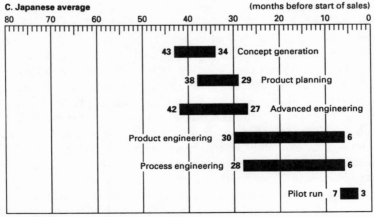

C. Japanese average

Note: average lead time of 12 Japanese projects

Figure 11.3 Average project schedule by stages.

184

Table 11.1. Ranking Based on Selected Indicators of Total Product Quality (TPQ)[a]

Region and strategy	Total quality ranking[b]		Conformance quality ranking[b]			Design quality ranking[b]					Overall rank	Value-adjusted overall rank	Customer base share change[c]	TPQ index[d]
	Consumer Reports (1)	Consumer Reports (2)	J.D. Powers (1985)	J.D. Powers (1985)	J.D. Powers (1987)	Concept	Styling	Performance	Comfort	Value for money				
Japan (volume producers)	●	●	●	●	●	●	○	●	●	●	●	●	●	100
	○	●	○	●	●	○	○	○	○	●	○	○		40
	●	●	●	●	●	●	●	●	○	●	●	●	●	80
	●	○	●	●	●	●	●	●	○	●	●	●	●	100
	○	○	n.a.	n.a.	○	○	○	○	○				○	25
	n.a.	n.a.	n.a.	n.a.	n.a.								○	23
		○	●	●	●	●	○	●	○	●	●	●		58
	○													35
United States (volume producers)	○	○	○	○	○	○	●	○	●	○	●	●		15
	○	○	○	●	○	○	●	○	●	○	●	●		24
				●		●	●	○		●				75
						●	●	○	○	●				75
	○					○	○		○	○	○	○		14
Europe (volume producers)	○	○	n.a.	n.a.	n.a.	○	○	○	●	○	○	○		47
	n.a.	n.a.	n.a.	n.a.	n.a.	●	●	●	●	○	○	○		39
			●			●	●	●	○		○	●		30
			○	○		●	○	○	●		●			35
	○	○		○	○	○	●	●	●	○	○	○		55
Europe (High-end specialists)	○	○	○		○	●		●	○		○	○	●	70
	●	●	●	●	●	○	●	●	●	●	●	●	●	73
	●	○	●	○	●	●	●	○	●	●	●	●	●	93
	●	○	●	●	●	○	○	●	●	●	●	●	●	100

[a]For further details of measurement, see Fujimoto (1989, chap. 5, Table 5.2).

[b]Entries in the map are defined as follows: Ranking in total quality (3 indicators), conformance quality (2 indicators), and design quality (7 indicators): ● = top one-third; ○ = middle one-third; none = bottom one-third.

[c]Customer base share changes based on four measures of long-term market share: ● = increase in customer base share; ○ = maintained share; none = decrease in customer base share.

[d]Summary index of ranking across all indicators of quality.

185

influencing engineers' implementation decisions; manufacturing capability affects the entire design process and, through this, impacts every aspect of design performance: cost, lead time, and quality.

To explore this relationship we examine four key stages in a design project's evolution: prototyping, the acquisition of dies, pilot production, and ramp-up. In each of these critical design-build-test cycles the "build" is literally a manufacturing process. Through these examples, the ability of manufacturing skills to enhance design is illustrated clearly.

PROTOTYPE AND DIE PRODUCTION [2]

In this section we focus on the beneficial impact manufacturing excellence can have on critical design activities. We select two high-profile activities: prototype production and the production of tools and dies. These activities are, in effect, "hidden manufacturing" embedded in the design process (see Clark and Fujimoto, 1991, chapter 8).

Prototype Production

Building prototypes is an obvious manufacturing activity that lies in the middle of the development process. Prototype shops, though very different from volume production plants, are still manufacturing operations. They employ manufacturing systems that fit small-volume production—general-purpose equipment, multi-skilled workers, "soft dies," manual forming and welding processes, and stationary assembly booths or very short, slow assembly lines. Runs from 50 to 100 engineering prototypes are typically assembled in two or three batches or generations, with design changes resulting from tests of one generation incorporated in the next.

The engineering prototype is the first physical object that represents the total product in terms of structure, material, look, and function. Earlier versions, such as clay models, mock- ups, and mechanical prototypes, represent the product only partially. A clay model may look like a real car but cannot move; a mechanical prototype may be able to reproduce the ride and handling of a new model but represents neither its exterior nor body structure. The engineering prototype provides the first opportunity to evaluate total vehicle performance and detect potential problems associated with the vehicle as a whole rather than with individual components.

Figure 11.4 compares regional averages of lead time to build a first prototype. Lead time, defined here, consists of the drawing release period and post-release period. The release of prototype drawings to prototype shops or prototype parts suppliers one after another rather than simultaneously accounts for the length of the drawing release period, which may reflect lead time for prototype parts procurement as well as for drawings. The post-release period represents the time between the release of the last drawing and completion of the first prototype.

The data show significant regional differences. Total prototype lead time in an average Japanese project is approximately six months, significantly shorter than that of the average U.S. and European project (approximately 12 and 11 months, respectively). The advantage of Japanese firms seems to come from differences in both the

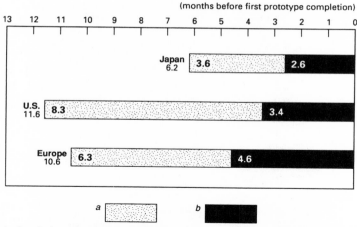

Figure 11.4. Lead time for first engineering prototype. Numbers do not add up exactly because some respondents reported total prototype lead time only.
[a]Drawing release (of first component to last component).
[b]Post-release (last drawing release to completion of first prototype).

engineering release process and in the way the prototype production process is managed. That the greatest variation occurs in the drawing release period suggests that control of parts procurement and the interaction of engineering changes and prototype projection drive the difference. Our interviews revealed that comparatively less attention was paid to time as a critical aspect of the prototype process. U.S. firms, in particular, suffer from a lack of scheduling discipline at prototype parts suppliers and in engineering drawing release. Moreover, information moves less quickly between design engineering and the prototype unit in U.S. firms. When product design changes during the prototype process in a typical Japanese company, design engineers go immediately to the prototype shop and instruct the shop technicians directly about the changes. They do not wait for paperwork. This contrasts sharply with the U.S. practice of subjecting design change orders to a series of formal approvals before allowing them to reach the prototype shop floor.

The kind of information generated and the kinds of problems solved by prototypes depends on the role of the prototype process. Our clinical and statistical evidence reveal two contrasting paradigms of prototype development: "prototype as early problem detector" and "prototype as master model." European high-end specialists view the engineering prototype as a master to be copied by the production model. In this master model paradigm, no time or expense is spared in ensuring the completeness and quality of the prototype. Production models are adapted to prototypes rather than vice versa. In short, this view of the perfect prototype and somewhat inferior production copy fits well with the strategy of high-end specialists, which emphasizes perfection of product functions uncompromised by cost and lead time.

In contrast, the early problem detector paradigm, followed by many Japanese carmakers, regards the prototype as a tool for finding and solving design and manufacturing problems at early stages of product development. The prototype is expected to anticipate the production model sufficiently that testing will reveal product and

Table 11.2. Characteristics and Paradigms of Prototype Fabrication

	Japan	United States	Europe (volume producer)	Europe (high-end)
Lead time	Short (6 mo.)	Long (12 mo.)	Long	Long (11 mo.)
Quantity	Mid-range (38/body)	Mid-range (34/body)		High (54/body)
Unit cost	Mid-range ($0.3 mil.)	Mid-range ($0.3 mil.)	Mid to high ($0.3–0.5)	High ($0.6 mil.)
Representativeness and design performance	Reasonably high representativeness and conformance	Sometimes low on both dimensions	Reasonably high on both dimensions	Very high on both dimensions
Prototype parts suppliers	Mostly production suppliers	Mostly prototype parts specialists	Mixed	Mixed
Paradigm	Prototype as problem detector (both product and process)	Prototype as a tool for proving product design	Split between master model and prove out of product design	Prototype as master model

process problems. Thus, the emphasis is on reasonably high representativeness, not perfection, of the prototype. Rapid construction of many prototypes, because it affords more opportunities to identify and remedy problems, is also important. The early problem detector prototype might be regarded as a draft of the production model, rather than the fully matured master of the master model paradigm.

Looking at Table 11.2, we see that the characteristics of Japanese prototype development, short lead time and reasonably high representativeness, are consistent with the early problem detector paradigm. Emphasis on rapid construction of prototypes and close communication between prototype shops and production facilities fosters early and accurate problem detection.

The quick turnaround of a series of prototypes while designs are still somewhat fluid means that Japanese prototypes are not perfectly representative of the final product. Our evidence suggests that prototype representativeness is greatest for European high-end specialists, followed by European and then Japanese volume producers. U.S. engineers frequently expressed dissatisfaction with the level of prototype representativeness, which has become a major concern of U.S.-based producers in recent years.

What Japanese producers lose in the representativeness of the model they gain back in the ways in which they use the model as a source of problem-solving for both product and process. The benefits of this early problem-solving effort are seen at the pilot stage when fewer pilots are produced with a manageable level of disruption of existing production.

The Japanese approach places a premium on extracting as much relevant information as possible out of each prototype. Regional differences in the use of prototypes shape and are shaped by regional differences in the supplier base. Japanese producers use primarily production suppliers rather than specialty prototype houses as prototype vendors. The two firms' close relationship and the production supplier's experience help to ensure maximum information is obtained.

Rapid turnaround of prototypes enables Japanese companies to look at a series of successive refinements. With longer lead times this approach could extend the total design process. Our evidence suggests that a one-month reduction in prototype lead time results in a one-month reduction in total engineering lead time. This result is consistent with the generally accepted notion that building the first prototype is on engineering's critical path.

Tools and Dies

Construction of tools and dies in preparation for volume production is another hidden manufacturing activity. Consider just the dies needed for body panel stamping. A typical car body is partitioned into some 100 to 150 stamped body parts. Each major panel may require as many as four or five dies, sometimes more if the design is complex. The number of dies produced for a project thus could range from several hundred to more than a thousand per body type. The number increases as the number of body types, stamping plants, and backup dies increases.

The process is similar whether in-house tool shops or outside die suppliers do the work. Dies are cast or forged, machined, finished, and assembled in a job shop setting by machinists using sophisticated general-purpose equipment. Although numerical control, computer-aided design, and other computer-based tools have had a major impact on manufacturing, an excellent match between upper and lower dies generally requires precision finishing by highly skilled workers. To meet the exacting requirements of the assemblers and accommodate the many engineering changes that typically arrive in the middle of production, die-making requires a substantial accumulation of technical skills and manufacturing know-how. It is a much more sophisticated and complicated process than many outsiders may imagine.

Building dies for body panels accounts for a large fraction of the total investment in a new car program and the greatest fraction of lead time in the engineering process. Consequently, outstanding performance in die manufacture may yield substantial advantages in overall performance.

Figure 11.5 compares overall lead time to design, manufacture, and test a set of dies for a body panel across regions. Die development begins with the first release of drawings for the body panel and continues through final release of the drawings and delivery of the manufactured dies. It ends with completion of the tryout of the dies in production. Though we see differences in lead time for each of the components of this process, the major gap between Japanese and Western producers occurs in the build phase—the period from final release to die delivery.

Why does die manufacture take 6 months in a Japanese tool and die shop and 14 to 16 months in U.S. and European shops? Advanced automation technology does not seem to be the answer. In fact, we found some U.S. and European producers equipped with high-tech die-making machine tools that we did not see in Japan. In any case, cutting time is an extremely small fraction of die-making time; as is often the case in manufacturing, what we must focus on to reduce throughput time is nonoperational time (e.g., downtime, inventory time). Our interviews and direct observations suggest that the Japanese advantage in die manufacturing lead time derives

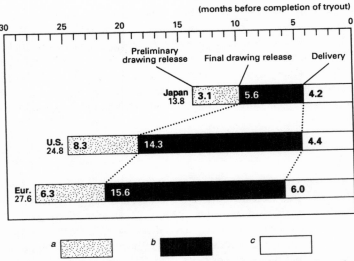

Figure 11.5. Lead time for a set of dies for a major body panel. Numbers do not add up exactly because some repondents reported total die lead time only.

[a]First to final drawing release for tooling order.

[b]Final drawing release to delivery of die (this approximately corresponds with die manufacturing lead time).

[c]Delivery to completion of tryout.

from overall patterns of die production, including management of engineering changes. Here we focus on two factors directly related to manufacturing capability: application of just-in-time (JIT) philosophy and integration of outside contractors.

Just-in-time in the tool and die shop. Japanese automakers have a long history of JIT in mass production and have disseminated much of this philosophy to their tool shops. To be sure, we did not see Kanban, Andon, U-shape line, or other tools typical of JIT mass production in Japanese die shops, yet the shops seemed to be strongly influenced by the JIT philosophy. For example, in Japanese die shops we saw many fewer work-in-process dies piled up in front of die cutting machines or in finishing areas than we did in U.S. and European shops.

The seemingly streamlined operations of Japanese die shops contrasts sharply with die making in U.S. and European shops, in which a conventional job shop philosophy prevails. Job shop managers oriented to maximizing utilization of expensive machine tools despite unpredictable and non-repetitive work flows tend to pile up buffer inventories in the hope that they will absorb job scheduling uncertainty. These high levels of work-in-process inventory have the effect of lengthening throughput time.

The supplier network. Differences in the die supplier networks may explain some of the Japanese advantage in lead time. U.S. carmakers have traditionally made arm's-length contracts with separate companies to carry out different manufacturing steps (e.g., molding suppliers for molds, casting specialists for casting, machine shops for cutting and finishing, jig suppliers for jigs). Such fragmentation makes it difficult to conduct die making steps in parallel and thereby compress manufacturing lead time. This is in contrast to Japan, where some major die suppliers offer as a package the entire die development process, including planning, design, and die manufacture. These suppliers may subcontract parts of the process, but the close, long-term relationships that prevail in the network enable them to integrate and overlap steps. Here, again, manufacturing capability in the supplier system contributes to higher performance in product development.

Manufacturing capabilities affect more than just the lead time of dies. Our research suggests that Japanese firms enjoy lower die costs as well, a significant advantage when die costs can represent half of a new model's capital investment (Clark and Fujimoto, 1991, p. 186). This cost savings stems from a variety of factors. A portion is due to supplier skills and the close supplier relations described above. The lower number of engineering change orders issued by Japanese designers also reduces die costs. Perhaps most illustrative of manufacturing's impact on design are the savings that stem from the ability of Japanese firms to perform fundamentally the same task with fewer dies.

A sheet of steel is transformed into a body panel through a series of stamping operations, each of which uses a different die (e.g., trim die, draw die, flange-up die, pierce die) to work the metal in a particular way so that, ultimately, the sheet is transformed into a complex shape. The number of dies required for a given panel is determined by the number of "shots" required to obtain the desired shape and properties (e.g., strength). Through continuous improvement of stamping practices, including operating practices, modifications to equipment, changes in the surface quality of steel, and attention to lubricants, better Japanese shops are able to use large transfer presses to stamp more complex dies and still achieve a high level of machine uptime and product quality. The result is that a typical Japanese body stamping plant needs only five shots (five dies and five tandem press machines) to make a complicated body panel (e.g., a quarter panel) that would require seven shots in a typical U.S. or European operation. A higher level of manufacturing quality—in this case, process control in commercial production—can yield a significant advantage in development productivity.

In die production we again see the power of manufacturing capability in producers and suppliers. In both lead time and die cost, an integrated network of highly capable tool and die manufacturers creates a significant advantage for Japanese automakers, which organize and manage their own internal operations to capitalize on the capability of these suppliers. The effect is a die manufacturing system that in the very best firms produces dies at half the cost of and in half the time as the U.S. and European systems.

As with prototypes, however, it is not just that Japanese firms are able to acquire dies more quickly and cheaply. In prototype production we noted that Japanese firms appeared to accomplish more problem-solving in the prototype phase. The same appears to be true in die production.

A comparison of Figures 11.3a–c and 11.5 brings to light an apparent anomaly. Japanese firms receive a complete set of dies in an average 13.8 months; U.S. and European firms take more than two years on average (24.8 months in the United States, 27.6 in Europe). But, when we look to Figure 11.3, process engineering, of which die production is a central task, takes 25 months on average for U.S. firms, 27 months on average for European firms, and 22 months on average for Japanese firms. What happened to the roughly one year of time saved due to die lead times?

Our evidence suggests that much of that time went into refinement of die designs. When we adjust the length of process engineering for differences in product content, scope, and engineering-manufacturing capability, we find that Japanese firms actually take more time in process engineering than their U.S. competitors. For the average car in our sample, we find that U.S. firms spend 21 months in process engineering (the average car in the sample is less complex than the average U.S. project), Japanese firms, 27 months. In general we find that capability for rapid prototyping and die development gives Japanese firms effectively six to ten months of additional time for process development and refinement.

Differences in level of refinement also show up in the performance and problems experienced during pilot production. The end of process engineering is marked by final sign-off on the design by the engineering organizations. It does not necessarily mean that all the tools and dies are complete, nor that there will be no further engineering changes. Indeed, our fieldwork suggests that it is quite common in U.S. projects for pilot production to begin with some parts produced by prototype tools and dies. Moreover, a large number of engineering changes occur after pilot, and even after ramp-up, has begun.

Evidence that U.S. product and process designs are much less refined when they reach the pilot stage is found in the number of prototype and pilot vehicles produced by Japanese firms. Though Japanese firms vary little from other volume producers in terms of number of prototypes per body style, they make far fewer pilot vehicles. This suggests that each prototype is a more powerful problem-solving tool and that product and process designs are consequently much more complete when they reach the pilot plant. The implication is that product and process engineering in U.S. projects (and European projects as well) spill over into pilot and ramp-up even though the designs have been formally released. In effect, U.S. and European firms failed to complete the design- build-test cycle as effectively as their Japanese competitors.

Thus, manufacturing capability at both the firms we studied and their suppliers makes it possible for Japanese firms to solve design problems more effectively. This effectiveness is illustrated by fewer defects remaining at pilot, more rapid supply of prototypes and dies, and lower cost dies.

PILOT RUNS AND RAMP-UP

The clearest instances of design-manufacturing interface occur during pilot and ramp-up production. When engineering has signed off on the design, when prototypes have been built and tested and the production tools and dies produced, all that remains is to bring everything together to see if it works as planned. The first step is to do a pilot run—a full-scale rehearsal of the commercial production system, including parts, tools, dies, and assembly. A successful pilot run is followed by ramp-up—the start-up of commercial production, which begins slowly and gradually accelerates to full production. The purpose of pilot runs is to find and solve problems that have gone undetected in prototype production and testing.

Problems also are expected to be found and solved during ramp-up. But ramp-up brings an additional management challenge. Because ramp-up is the initial phase of commercial production, it is vital that it adhere to schedule and reach full volume as quickly as possible. Delays translate into a later return on the firm's investment in design or, worse, permanently lost sales. Yet rapid acceleration of production can create confusion and often leads to quality problems. Because these quality problems occur in the first vehicles to reach the market, they can have a critical impact on market perception and economic success. Defects and poor reliability discovered during a product's introduction period can permanently destroy its reputation and image. Press and word-of-mouth reviews of a new product tend to focus on recurring defects such as squeaks and rattles, poor paint quality, and misaligned trim. To protect their product's reputation and ensure future sales, automakers go to great lengths to ensure that only high-quality cars are introduced to the market.

Thus carmakers are caught between trying to bring production volumes up as quickly as possible while finding any and all defects as quickly as possible. These dual forces place a premium on a firm's manufacturing capabilities. Quick response, rapid problem-solving, volume flexibility, and an ability to manage in a rapidly changing environment are all critical assets.

But the success of pilot and ramp-up is dependent only in part on the excellence of manufacturing. It is also contingent on the quality and completeness of the design prior to pilot production and the effectiveness of the engineer in quickly correcting errors found in pilot production before ramp-up begins. The diversity we find internationally in pilot and ramp-up experiences is rooted in differing manufacturing and design capabilities.

Pilot Runs

Final assembly is one of the most extensive, yet least automated, processes in modern automobile manufacturing. It typically consists of a main assembly line of a few hundred stations, complicated branches of subassembly lines, thousands of parts boxes alongside the line, and hundreds of assembly workers per shift. The challenge for the pilot run is to simulate this complex process accurately and train workers properly with minimal sacrifice in cost and schedule.

Pilot run schedules are shown in Figure 11.6 for three U.S., four Japanese, and three European development projects. Black bars indicate ramp-up of commercial production. Pilot runs are differentiated according to where they are carried out: at separate pilot plants (blank bars); on separate pilot lines within volume plants (lightly shaded bars); or on volume production lines (darkly shaded bars). The figure indicates that several pilot runs tend to be conducted, first on separate pilot lines or in pilot plants and later on production assembly lines. The fraction of real production tools used, and thus the reality of the rehearsal, is greater in the later pilots. Except for the US1 and US3 projects, in which the plant was totally renovated for the new car pilot, existing lines with minor modifications were used to pilot the new models.

Figure 11.6 shows significant differences in the pattern of pilot runs, even within regions. Among Japanese producers, for example, the number, length, interval, and location of runs vary widely. But relative to U.S. and European projects, Japanese projects have a number of characteristics in common:

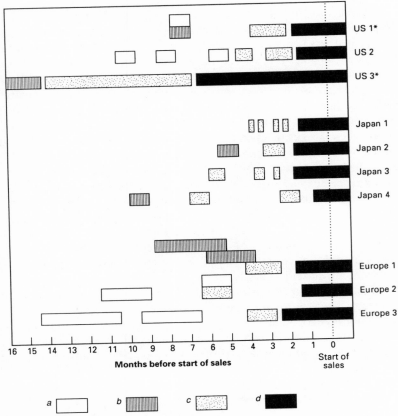

Figure 11.6. Schedule of pilot runs for selected projects.

*Major renovation of assembly line; otherwise modification of existing assembly line.

[a]In a separate pilot plant.

[b]On a separate line within volume plant.

[c]On an existing volume line.

[d]Volume production (ramp up).

- Pilot runs are relatively short
- Pilot run periods are compressed
- Pilots are more often carried out on production assembly lines

The use of actual production lines to produce pilot vehicles simultaneously improves the fidelity of the experiment and begins to train members of the local work force to produce the new car. In this way it serves the purposes of pilot runs better than any other approach. In addition, this approach saves the capital that would be invested in a pilot site. The tendency of U.S. and European firms to locate pilot production in a separate facility appears to lengthen the problem-solving cycle and complicate knowledge transfer from pilot to commercial production. None of the Japanese firms we studied operates separate pilot plants; pilot runs start in the main assembly plant, either on a small pilot line surrounded by walls or curtains for confidentiality or on the existing assembly lines.

The price for this fidelity is the disruption of existing production. The newness of some pilot manufacturing tasks and the expected problems that will be found mean that the pilot vehicles' cycle time at each assembly station will exceed that of the production vehicles on the line. The simpler solution, to shut down commercial production and reduce line speed for the pilot, is associated with significant production loss. The other solution is to mix current models with pilot vehicles; workers assemble new and existing models at the same station, at the same line speed, and with similar task assignments, using "empty hangers" to absorb differences in productivity. Typically, a pilot vehicle is preceded and followed by two vacant body carriers, which allows workers to spend five minutes instead of one on the unfamiliar model.

The "empty hanger" approach, being an application of mixed- model assembly, inevitably complicates work assignments and part handling. Nevertheless, the approach minimizes the opportunity cost for the pilot run by reducing lost production of the existing model. More importantly, the approach involves future workers in early training in a very realistic setting, which facilitates early problem detection by these workers and their supervisors. In addition, as one Japanese process engineer pointed out to us, workers tend to get excited about a new model they see on the line and become motivated to learn more about it.

Simultaneous execution of the pilot runs and commercial production on the same line, though advantageous, requires a high level of manufacturing capability, including discipline and clarity in material handling and production planning, skilled workers and supervisors, and process control that provides the flexibility needed to cope with the complexity that arises out of the mix of new and existing models. In sum, it relies on superior manufacturing capability. Japanese producers follow this approach because their manufacturing plants can do so at reasonable cost.

While superior manufacturing capability is a necessary condition for effective pilot production on a commercial mixed-model line, it is not sufficient. The design also must be reasonably complete and accurate. In mixed-model production, empty hangers provide a temporal buffer around the vehicle. But if the task cannot be completed in the allotted time, line imbalances will either force production to slow or leave tasks incomplete. The former raises costs since it slows commercial production. The latter reduces the fidelity of the pilot experience. Either commercial needs or design needs must give if tasks exceed the time available. Significant assembly problems, or at the extreme, parts that cannot be assembled due to misaligned holes, incorrect dimensions, or other design errors, wreak havoc with mixed-model assembly on commercial lines, placing a premium on the type of problem-solving discussed earlier. These problems are far less damaging on flexible pilot lines with no competing commercial production.

There is reason to believe that the Japanese preference for capital-conserving, high-fidelity pilot production on commercial lines may stem in part from the higher degree of certainty regarding the nature of design problems remaining at the pilot stage. In fact, Japanese firms produce fewer pilot vehicles per body type than manufacturers in other regions. Japanese producers average 53 pilot vehicles per body type, U.S. producers 129, and European producers 109 for high-volume producers and 205 for high- end producers. This may be due in part to the higher fidelity of the pilot production environment, but differences in design practice also appear to be important.

Table 11.3. Ramp-Up Performance by Region[a]

	Japan	United States	Europe
Plant shut-down period (mo.)	.25	10[b]	.5
Time from start to full production (mo.)	3	5	5
Time from production start to sales start (mo.)	1	4	2
Time to return to normal productivity (mo.)	4	5	12
Time to return to normal quality (mo.)	1.4	11	12

[a] 6 Japanese, 3 U.S., and 3 European projects.
[b] All U.S. cases involved major plant renovations.

For some producers, superior manufacturing capability enables design to conduct a more realistic "dress rehearsal" in pilot production: a clear case of benefiting from design's manufacturing excellence. But manufacturing capability alone will not make this approach cost effective. For the production line to operate smoothly there must be no egregious errors in design. The design need not be perfect, but it must be reasonably complete and accurate. Here design excellence benefits manufacturing. A similar symbiotic relationship exists in the first critical task of what is generally regarded as true automobile production: ramp-up.

Ramp-Up in Assembly

As Table 11.3 indicates, patterns of ramp-up performance varied widely across regions. In part this is due to differences in tasks; two of the U.S. plants were bringing new lines up as well as new products, for example. Even attempting to control for these differences, however, the superior performance of Japanese projects is striking.

Just as striking as the differences in ramp-up performance are differences in ramp-up practices. Ramp-up procedures can be differentiated along two key dimensions: ramp-up curve and work force policies. The primary choices are shown in Figure 11.7, which illustrates three basic patterns of ramp-up curves. In shutdowns, the old model is terminated completely, and then, normally after a short interval, the new model starts up (see Figure 11.7, panel 1a). This is a relatively simple approach in that no mixed-model production between the old and new models is required. Loss of production and sales, however, is potentially high. To reduce loss, the company must have a very steep ramp-up curve.

In block introductions, different versions of a model are introduced with significant intervals, typically six months to one year (e.g., domestic first, then export or sedan first, then wagon). For the production line this means that ramp-up occurs in two or three blocks (see Figure 11.7, panel 1b). Mixed-model production during the transition period may add complexity to the operation, but the problem of lost sales and/or a steep ramp-up curve is somewhat alleviated because the old model serves as a cushion. In step-by-step introductions, an extreme form of block introductions, the

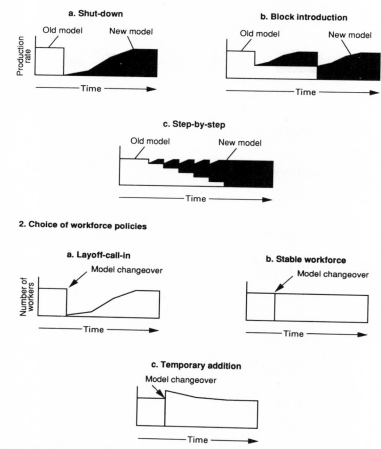

Figure 11.7. Choices in production start-up.

old model is phased out while the new model is gradually phased in, creating many small, sawtooth ramp-up curves (see Figure 11.7, panel 1c). Although lost sales are kept to a minimum, a smooth transition requires subtle and continuous adjustment of material handling, work assignments, and scheduling. All three approaches—shut-down, block introductions, and step-by-step—benefit from manufacturing excellence. But what aspects of excellence are critical varies.

Figure 11.7 also illustrates three different approaches to the hiring of workers and their job assignments. In layoff-call-in (see Figure 11.7, panel 2a), most of the workers on an existing line are temporarily laid off before start-up. As the production rate rises, they are gradually called in. If the plant is newly built or renovated, workers are simply called in step-by-step. In this approach, job assignments change rapidly during start-up: workers at the initial stage must handle a very broad range of tasks, which narrows as tasks are reassigned to new workers added to the line. The stable work force approach (see Figure 11.7, panel 2b) leaves the number of

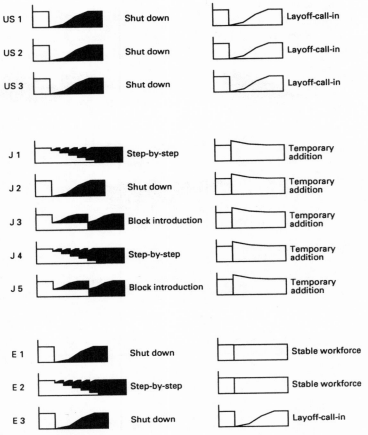

Figure 11.8. Ramp-up choices.

workers basically unchanged before and after the model changeover. Compared to the layoff approach, a greater labor cost may be insured to maintain the work force, but work force continuity is high. There are some cases (e.g., German producers) in which companies are forced to adopt this policy because of regulations restricting layoffs. The temporary addition approach (see Figure 11.7, panel 2c) actually increases the number of workers during model changeover, typically by about 10%, and then drops back to normal staffing levels when production is normalized.

The ramp-up choices made at each site studied are reported in Figure 11.8. The greatest consistency is shown in U.S. sites where shutdown/layoff-call-in dominates, in part due to the major renovation of two of those plants. The shutdown approach was also the dominant choice in Europe, although greater variation in work force policies existed (possibly due to local regulation of layoffs). Japanese plants exhibit tremendous diversity in their approaches to the introduction of volume but consistency in their work force policies. These results suggest that there is no one best paradigm for ramp-up.

It is our hypothesis, however, that the patterns shown reflect something more than simply differing choices. Plants that face the likely prospect of major defects encounter a difficult ramp-up problem. Line balance is virtually impossible early on, whether it is attempted through adjustments in line speed, empty hangers, or some combination thereof. Line balance and staffing must be readjusted contingent on the problems that surface.

In this situation, the shutdown/layoff-call-in ramp-up is the lowest-risk approach. Production levels can be adjusted to reflect current problems without affecting production of other vehicles. In either block introduction or step-by-step introductions an increase in new vehicle processing time translates into reduction in the production of the older vehicle, thereby raising costs. Similarly, the layoff-call-in as needed approach minimizes the cost of line shutdown by reducing the number of people who will be idled. Layoff-call-in is, like off-line pilot production, well suited to environments in which a great deal of uncertainty exists regarding design quality.

The types of defects that appeared during ramp-up differed significantly by region. Typical start-up quality problems in Japanese plants were relatively minor; scratches, leaks, and noise were all cited. U.S. plants typically experienced far more significant problems—misaligned bolt holes, for example, which caused the line to shut down production. European plants experienced problems in between the U.S. and Japanese plants in terms of severity; wrong parts, for example, might still shut down the line but did not require engineering redesigns to solve. Reflecting these different problem types, the companies also differed in who took primary responsibility for solving start-up problems. In Japanese firms, line supervisors played the key role; elsewhere, engineers were the key problem-solvers.

This is not meant to imply that ramp-up is an easy task where major defects are few. The success of Japanese sites in ramping up new products is only partly due to design excellence. It is also facilitated by their greater use of common parts from one product generation to the next (see Clark, Chew, and Fujimoto, 1987). The use of extra workers during early ramp-up also gives the production line a greater ability to respond, within limits, to a few unexpected problems by assigning extra staff to affected workstations.

Manufacturing capability plays a critical role as well, especially with regard to superior materials coordination (a problem in product ramp-up in many European sites) and quicker shop floor response to problems. A manufacturing engineer in one Japanese company, for example, pointed out that manufacturing and plant engineers are the core problem-solvers at the pilot stage but that supervisors become the main players as commercial production begins. This implies that serious defects requiring significant engineering resources are mostly solved prior to start-up—a benefit of early problem-solving noted earlier. Interestingly, the famous quality circle, suggestion system, and other employee-involvement programs are not emphasized during the ramp-up period. As a manufacturing engineer in a Japanese company suggested, quality circles may be adequate for continuous improvement in regular periods, but they are too slow in making decisions in ''war times'' (such as ramp-up). Line supervisors are in a better position to lead problem-solving in ramp-up partly because they are themselves experienced technicians and partly because they are on the spot when problems occur. Supervisors also may have received additional training in early pilots. Thus, as an engineer points out, supervisors have ''good eyes'' to identify start-up problems quickly and effectively. Their ''good eyes'' are supplemented by good information. In one plant of one of the better Japanese producers, for example, information on defects is announced to the entire plant floor through in-house broadcasting systems as soon as they are detected.

Ramp-up performance, like pilot production, illustrates the importance of both manufacturing and design excellence. Before leaving the issue of ramp-up, it is worth touching on the large variation in choice of ramp-up curves seen in Japan. We believe that these too reflect firm-specific manufacturing capabilities.

Ramp-up, as noted at the outset of this chapter, is an inherently difficult task. There are two key problems: changing over from commercial production of one vehicle to the next and rapidly increasing new vehicle volumes. The different ramp-up approaches place different emphasis on which key problem will be most difficult. In the shutdown approach, the problem of changing over from one vehicle to the next is greatly reduced. However, the difficulty and importance of rapidly increasing volume are increased. Step-by-step production does the opposite: volume changes are more restrained, but workers are continually faced with the problem of changing from old to new vehicles.

Which choice a firm makes appears to be predicated on the existing strategy for volume production of product. The companies that have emphasized mixed-model production on a single line, adjusting volume by changing mix, find the step-by-step approach a manageable challenge for both workers and managers. Companies that have emphasized dedicated production lines (one vehicle per line) and have responded to volume changes by slowing down and speeding up the line find shutdown the more manageable approach. Neither approach is intrinsically superior. It is a question of fit with manufacturing capabilities. Reflecting the importance of fit, two of the better Japanese producers—one using a shutdown approach and the other step-by-step—explained separately that they couldn't do their competitor's alternative approach because "it is much too hard."

Beyond the importance of manufacturing and design capability for a smooth ramp-up, which is crucial for market success, manufacturing capability plays another significant role. The certainty that design possesses regarding manufacturing's ability to ramp-up to high-quality volume production quickly (3 months for Japanese firms and roughly a year for U.S. and European firms) allows design to organize its process differently. Product engineering is completed 12 months in advance of commercial sales by U.S. firms in our study, as shown in Figure 11.3. European firms freeze the product's design a full 19 months in advance. Japanese firms freeze the design only 6 months in advance. One must have a tremendous amount of confidence in manufacturing's ability to ramp-up production with that little room to maneuver. Product planning tells a similar story, beginning in the average Japanese project after the completion of product planning in U.S. and European projects.

This ability to work closer to the market is perhaps the most important influence that manufacturing capability brings. The ability to execute build tasks well and solve problems quickly means that Japanese producers can make product decisions at a point in time much closer to launch. In this way they increase their chances of market success. Their ability to start development closer to the market is due in part to the firms' ability to coordinate across the stages in the design process: in design capability. It also rests on the firms' ability to complete prototype, die production, pilot production, and ramp-up quickly and effectively. As we have shown, these tasks rely heavily on manufacturing capability.

BEYOND THE PRODUCTION/R&D DICHOTOMY

We have looked at four key events in a new product's life: pilot production, ramp-up, prototype production, and the production of dies. For each, we have attempted to show that design skills and manufacturing skills can build on each other to achieve significant competitive advantage for the firm that masters both sets of skills. The best firms are neither design for manufacturing nor manufacturing for design; they are designing and manufacturing for competitiveness.

Before setting guidelines for resolving the design/manufacturing conflict, one would do well to ask why the conflict exists at all. In truth it is more a designer/manufacturer conflict than it is a design/manufacturing conflict, i.e., it is rooted more in the organization than it is in the design and manufacturing tasks themselves. Our research on the world auto industry suggests that design and manufacturing skills have critical elements in common.

Design and manufacturing tasks have traditionally been thought of as distinct, radically different activities. This perspective stems in part from differences in the two physical environments. Standing in the body shop of an automobile plant highlights the striking differences between the early stages of design and the production of product for sale. Huge, multi-million-dollar presses noisily stamp out body panel after body panel from sheets of steel. Numerous robots whir and dip and weld. Amidst this noise, regimented workers monitor and troubleshoot, load and unload.

Contrast this with the high-ceilinged design studio, with its large windows and tasteful, pristine working areas. Designers sit at large drafting boards sketching out new concepts that model shop workers—as much artists as employees—render in clay, plastic, or fiberglass downstairs.

Such sharp differences in environment might seem to suggest that the management of design and of manufacturing are quite different disciplines—the former directed at inspiring creativity, the latter at meeting quotas. This perspective is reinforced by a body of academic research that often stresses the dichotomy between R&D and production.

But this research, like our choice of work environments above, has stressed opposite ends of the R&D-manufacturing spectrum. R&D studies have tended to center on ''R''—on laboratories carrying out basic research. When commercial product development is studied, it is often of a unique and/or breakthrough kind: for example, there are many studies of major defense contractors or the more limited work on ethical pharmaceuticals. Naturally these studies have focused on the most fluid and creative aspects of R&D.

Similarly, manufacturing researchers as far back as Taylor and Gilbreth have focused on the most stable and repetitive of manufacturing tasks. By examining production in a single, unchanging product/constant technology environment like that of Taylor's carefully detailed hod carriers, one naturally tends to highlight control, stability, and conformance to standard practice as principal management goals.

But for the design and manufacture of most commercial products this dichotomy is a false one. In essence, the design process produces a product much as any other process produces a product. The design product differs only in that it is an "information asset" rather than a physical asset. Design is a predictable, repetitive process in the sense that each design will proceed through a set of predetermined, sequential processing steps. The form of the information asset changes as it progresses from upstream to downstream in the design process. The design moves from sketches to drawings to models to CAD/CAM database to prototype to pilot run to production. At each stage its form is modified and value is added. In fact, we have shown that the production of designs contains steps that are not simply analogous to manufacturing but are literally manufacturing activities, albeit in low volumes.

Manufacturing processes are virtually the same, but the emphasis is reversed. The manufacturing process's principal output is the physical asset, the cars themselves. But along the way the manufacturing process also engages in a variety of information- generating activities. Troubleshooting, quality circles, employee suggestions, and pilot production are all information-generating events. The whole of total quality control can be thought of as emphasizing the information-generating role of the production process.

Both the manufacturing process and the design process have outputs of two types: physical and information. The physical outputs differ in their use. Design's physical outputs are used in-house or by suppliers, either by design groups (as in the case of prototypes) or by manufacturing groups (as in the case of tools and dies). The physical output of manufacturing is, of course, sold to consumers.

Information assets created by the two groups are used far more similarly. Design generates information, through prototype testing, for example, that it uses further to refine and modify the design to enable it to better achieve its goals. Manufacturing generates information, through statistical process control, for example, which it uses further to refine and modify the production process to enable it to better achieve its goals.

When described in these terms the R&D-manufacturing dichotomy becomes far less stark. One could easily believe that these functions would have much in common despite their differences in production volumes, locations, and titles. Recently, researchers and practitioners have begun to speak of a new model of production, one built on the just-in-time and total-quality-control paradigms of Japan. Similarly, a new model for rapid product development has also recently emerged; it too first came to attention when researchers examined the origins of Japan's competitive success. Both paradigms are rooted in a similar philosophy, one that looks to cut across functional areas, solve problems rapidly, prevent errors, and continuously improve. The striking parallels in these two paradigms are summarized in Table 11.4.

The popular image of new car development as an activity that goes on in design studios and on test tracks overlooks the critical manufacturing activities embedded in the development process. Because these activities, though different from commercial production in scale and configuration, nevertheless involve the manufacture of physical objects, a company characterized by excellence in manufacturing may enjoy an advantage in product development. Similarly, excellence in product development di-

Table 11.4. Similarity of New Paradigms in Production and Development

Production (JIT-TQC paradigm)	Development (new paradigm)
Process-flow patterns	
Frequent set-up changes	Frequent model changes
Short production throughput time	Short development lead time
Reduction of work-in-process inventory between production steps	Reduction of informational inventory between product development steps
Piece-by-piece (not batch) transfer of parts from upstream to downstream	Frequent (not batch) transmission of preliminary information from upstream to downstream
Quick feedback of information on downstream problems	Early feedback of information on potential downstream problems
Quick problem solving in manufacturing	Quick problem solving in engineering
Upstream activities are triggered by real time downstream demand (pull system)	Upstream activities are motivated by downstream market introduction date
Organizational capability	
Simultaneous improvement in quality, delivery, and manufacturing productivity	Simultaneous improvement in quality, lead time, and development productivity
Capability of upstream process to produce saleable products the first time	Capability of development (i.e., upstream) to produce manufacturable products the first time
Flexibility to changes in volume, product mix, product design, etc.	Flexibility to changes in product design, schedule, cost target, etc.
Broad task assignment of workers for higher productivity	Broad task assignment of engineers for higher productivity
Attitude and capability for continuous improvement and quick problem solving	Attitude and capability for frequent incremental innovations
Reduction of inventory (slack resources) forces more information flows for problem solving and improvements	Reduction of lead time (slack resources) forces more information flows across stages for integrated problem solving

rectly benefits the manufacturing group with enough skills to exploit new opportunities.

It is our contention that firms should look to instilling the managerial approaches that build both design and manufacturing capabilities to achieve long-run competitive vigor. Achieving the integration of manufacturing and design is important, but more important is the fundamental task of creating new, world-class capabilities in both groups.

ACKNOWLEDGMENT

This research was supported by the Division of Research, Harvard Business School.

NOTES

1. Data in this section were developed as part of a larger study of product development in the world auto industry. For a review of that study and more extensive analysis, see Clark and Fujimoto (1991).

2. The material in this section is adapted from Chapter 8 of *Product Development Performance*.

REFERENCES

Clark, K. B. "Project Scope and Product Performance: The Effects of Parts Strategy and Supplier Involvement on Product Development." *Management Science* 7(10), 1989, 1247–1263.

———, W. B. Chew, and T. Fujimoto. "Product Development in the World Auto Industry." *Brookings Papers on Economic Activity* 3, 1988, 729–771.

——— and T. Fujimoto. *Product Development Performance: Strategy, Management, and Organization in the World Auto Industry.* Boston: Harvard Business School Press, 1991.

Fujimoto, T. "Organizations for Effective Product Development: The Case of the Global Automobile Industry." Unpublished Ph.D. dissertation, Harvard University, 1989.

III

SOCIAL, POLITICAL, AND CULTURAL CONTEXT

12

DEVELOPMENT OF A MODEL FOR PREDICTING DESIGN FOR MANUFACTURABILITY EFFECTIVENESS

GERALD I. SUSMAN and JAMES W. DEAN, JR.

This chapter reports on the development of a model to predict the success of efforts by design and manufacturing to work together on new product development projects. One source for development of the model was interviews conducted with personnel from 12 companies that were undertaking such efforts. Unstructured interviews were conducted initially with senior managers from the 12 companies during the spring of 1989, and then structured interviews were conducted with 129 persons from 21 projects during the summer and fall.[1]

Another source for development of the model was the existing literature on cross-functional coordination. Our approach to developing the model was to alternate regularly between the perspective gained from the interviews and that gained from existing theory and research. Each perspective informed and guided the other as the model was developed. That is, the interview data were organized into inductively derived categories and existing theory and research led to questions to ask company personnel during interviews. The chapter has been organized to reflect the natural evolution of the model.

The interview data were collected for the purpose of developing the model, not for testing it. As such, their presentation should be viewed as illustrative of a particular construct of the model or descriptive of one of the organizational policies or practices to which the construct relates. No one company necessarily exhibited all of the policies or practices that make up a construct. The model described below, therefore, represents a composite of best policies and practices used by companies to improve how design and manufacturing work together during the development of new products. Also included in the model are project outcomes that these policies and practices are assumed to impact favorably as well as variables that are assumed to moderate these relationships.

DESIGN FOR MANUFACTURABILITY

We use the term design for manufacturability (DFM) to characterize what these 12 companies were trying to do. Common to all 23 projects studied was that manufacturing was involved earlier in the product development process than it had been previously; that is, earlier relative to the types of products that each company has had experience developing. For example, some companies have been developing products that are so complex or have used materials that are so little understood that manufacturing cannot make meaningful contributions in the early phases of the product design. Thus, early involvement in companies that produce these products might seem relatively late for companies that produce simpler products or use materials that are well understood.

Projects also can vary by what manufacturing attempts to do when it is involved earlier. Manufacturing can inform design about its existing manufacturing capabilities so that design can take these capabilities into account in the product design, e.g., machining tolerances. Taking such capabilities into account during the product design phase reduces the chance that designers will have to modify the design late in the product development cycle. This approach is conservative in the sense that designers are asked to adjust to existing factory capabilities. The benefits are derived from fewer mismatches between product characteristics and existing process capabilities that are due primarily to the designer's ignorance of existing factory capabilities (Langowitz, 1988).

Manufacturing also can suggest ways to design the product for ease of manufacturability. This approach is more ambitious than the one described above because manufacturing can do more than just request designs that use existing process capabilities. They can suggest how products can be designed with fewer parts, assembled or tested more easily, or accommodated to automated equipment (Boothroyd and Dewhurst, 1988). In this case, the benefits are derived from a product design that is more economical to manufacture rather than from avoidance of mismatches between the product design and existing factory capabilities.

In either of these uses of early manufacturing involvement, manufacturing personnel need not necessarily be physically present in order to provide information such as machining tolerances and materials availability. They can be involved early by proxy if data about existing manufacturing capabilities or design for manufacturability can be codified and made readily accessible to designers while they are designing the product. This is likely to be possible only for mature products or for enhancements to existing products.

A third use of early manufacturing involvement is possible if manufacturing can learn enough about the product to design the manufacturing process while the product is still being developed. This approach is the most ambitious of the three in that the phases of the product development cycle overlap. Adjacent phases can overlap at their interface, as shown in Figure 12.1a, or the overlapping can extend across several phases, as shown in Figure 12.1b. Imai, Nonaka, and Takeuchi (1985) described the latter type of overlapping as characterizing the product development process at two major Japanese firms.

During overlapped phases, the product and manufacturing processes are being designed simultaneously. The more successful the communication between design and manufacturing during the overlap, the more the product and manufacturing processes can be designed simultaneously. In fact, the communication may be so successful that the product and process design may be completed at the same time. This is accomplished, in part, by making final product design decisions later in the development cycle than usual and making initial manufacturing process decisions earlier

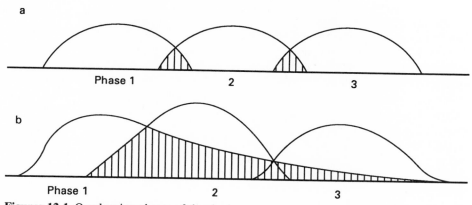

Figures 12.1 Overlapping phases of development.
Source: Adapted from Imai, K., I. Nonaka, and H. Takeuchi. "Managing the New Product Development Process: How Japanese Companies Learn and Unlearn." In *The Uneasy Alliance,* eds. K. B. Clark, R. H. Hayes, and C. Lorenz. Cambridge, Mass.: Harvard University Press, 1985, 337–381.

in the cycle than usual (Imai, Nonaka, and Takeuchi, 1985; Clark and Fujimoto, 1989). The benefits derived from this type of communication go beyond avoiding mismatches and designing an improved product and process. Time is saved by carrying out tasks simultaneously rather than sequentially.

Early manufacturing involvement augments the typical flow of communication between design and manufacturing. With each increasing level of ambitiousness, manufacturing increases the number of constraints that designers need to consider in their product designs. The flow of communication becomes more frequent and two way because design has to consider the plans of manufacturing in its product design. The information that each offers becomes more tentative and provisional because it is contingent on what the other party does with the information. We will discuss the implications of these changes in information properties and flow for the relationship between design and manufacturing in a later section of the chapter.

Reference only to design and manufacturing in discussions of design for manufacturability should not obscure the fact that each function contains a number of subfunctions. The design function includes personnel who design the product and/or test to ensure that it has met its functional requirements. The manufacturing function includes materials personnel, who want designers to use standard parts and materials that are readily available; quality- assurance personnel, who want designers to design products so that they are easier to inspect and test; equipment services personnel, who want to know as soon as possible if new machines need to be purchased; methods personnel, who want as few new manufacturing steps as possible to result from the design; and manufacturing personnel, who want the product to have fewer parts as well as parts that are easy to fabricate or assemble.

DESIGN FOR MANUFACTURABILITY EFFECTIVENESS

The 12 companies that participated in the study undertook DFM initiatives in order to increase the likelihood of achieving one or more project goals. These project goals were cost, quality, performance, and meeting schedule. Typically, success by DFM in facilitating one goal generally affects one or more of the other goals favorably.

DFM can reduce development costs by reducing the number of engineering hours required to correct mismatches between the product and process design. It also can reduce unit costs by leading to the redesign of products so that their parts are easier to fabricate or assemble. DFM also can reduce life-cycle costs by leading to the design of reliable products. Life-cycle costs also can be reduced by designing products that are relatively easy to maintain and service.

DFM can improve quality by leading to a better match between the product design and process capabilities, e.g., designing the product to existing machine tolerances, or by selecting parts that are more reliable. Quality, in this sense, refers to products that conform to standards or are reliable, i.e., high mean time between failure (Garvin, 1984).

DFM can improve product performance, which, for example in the auto industry, refers to product capabilities and features such as speed, weight, and fuel consumption. Marketing plays a critical role in determining what capabilities and features customers expect or demand. Design and manufacturing can contribute also through the use of quality function deployment (Hauser and Clausing, 1988) to achieve agreement with marketing on customer expectations and demands and to design a product that meets or exceeds them.

DFM can reduce lead time from design conception to delivery of the product. Lead time can be reduced by reducing or eliminating the time required to complete engineering changes issued after the product design has been released to manufacturing. Any DFM activity that improves quality by improving first-time yield and the time needed to reach an acceptable product yield also reduces lead time. Any improvement in communication between design and manufacturing that leads to simultaneous engineering also reduces lead time and improves the chances of completing a project on time or ahead of schedule.

The projects in our study closely ranked meeting schedule, performance, and quality as first, second, and third in importance, respectively, and ranked cost last. The reason the first three goals were ranked above cost may be due to increasing pressure from competitors or customers to deliver high-quality, high-performance products more quickly. For example, many foreign and domestic companies have learned how to get products to market earlier, thus putting pressure on their competition to do the same. Products in some industries become obsolete very quickly (Qualls, Olshavsky, and Michaels, 1981). Some companies may have a conscious strategy of accelerating product obsolescence by introducing new and modified products to the market frequently (Susman and Dean, 1989). They may recognize that the enhanced profits from getting a new product to market early is usually well worth the cost incurred in doing so (Reinertsen, 1983; Stalk and Hout, 1990). Some of the defense contractors in our study mentioned that they were being pressed by the Department of Defense to shorten their product development cycle and to improve quality.

Achieving one or more of the above four goals is important for competitive advantage, but some companies may consider it important to track progress on these goals by direct measures of DFM effectiveness. Such direct measures include the number of engineering change notices, first-time yield, ramp-up time, or rework as a percentage of direct labor. Success on these measures should contribute to achievement of one or more of the project goals, while more directly reflecting the consequences of initiatives aimed at improving communication and coordination between design and manufacturing. Our interviews indicated that few companies collected systematic data on such measures. Thus, any future effort to test our model of DFM effectiveness most likely will have to do so with data on achievement of project goals.

INITIAL CONDITIONS FACED BY DESIGN AND MANUFACTURING PERSONNEL

We earlier described DFM efforts as attempts to modify the pattern of interaction and flow of information between design and manufacturing during the product development process. In order for such attempts to be successful, it is necessary to understand the conditions that design and manufacturing face prior to introducing DFM activities.

Interdependence

Design and manufacturing are interdependent to the extent that each is constrained by the decisions or actions of the other or has information that the other needs to meet its specific responsibilities. The efforts of each project to involve manufacturing earlier in the design process are aimed at altering the typical asymmetry in communication and interaction between the two functions.

In the typical relationship between design and manufacturing, design constrains the options of manufacturing by designing the product without input from manufacturing. Early manufacturing involvement reduces the asymmetry between design and manufacturing by informing design about its capabilities or suggesting ways to design the product for ease of manufacturability. In these two cases, design still acts, while manufacturing informs or suggests. The asymmetry is reduced further in simultaneous engineering because both design and manufacturing are acting at the same time and informing and mutually constraining each other.

Interdependence between design and manufacturing typically is viewed as being sequential (Thompson, 1967). Design completes the product design and hands it over to manufacturing where the production process is then designed. This characterization of interdependence is not entirely accurate, however, because manufacturing often has to ask design late in the cycle to modify the product so that it can be produced to meet specific cost or quality goals. This occurs because design and manufacturing are reciprocally interdependent; that is, the product cannot achieve all of the goals set for it without inputs from both parties. The later in the product design cycle that design and manufacturing deal with reciprocal interdependence, the more expensive its consequences for modifying the product design.

Design and manufacturing need not necessarily deal with reciprocal interdependence if project goals can be lowered enough to permit design and manufacturing to forego interaction (Galbraith, 1973). This option is increasingly unlikely to exist in today's competitive marketplace, thus making the issue not whether design and manufacturing are going to deal with reciprocal interdependence, but where in the product development cycle they are going to deal with it. When they deal with it late, either the company or the consumer bears the burden of high cost and long delays. As companies now are less able to pass on such costs to the consumer, they are being forced to deal with reciprocal interdependence earlier in the product development cycle.

Reciprocal interdependence increases the frequency of communication between the interdependent parties, if for no other reason than that two parties are initiating communication instead of one. The overall frequency of communication will increase when the parties need to exchange information that is ambiguous or uncertain. The frequency of communication per unit of time will increase when lead time for delivery of the product is reduced. The information can be exchanged by computer network or electronic mail or within a small group or in an ad hoc one-on-one setting. The conditions influencing the appropriate medium and setting will be discussed later in the chapter.

Reciprocally interdependent parties also can vary in the degree to which their respective actions have positive or negative outcomes for the other. McCann and Galbraith (1981) use the term externality to discuss the impact of the decisions of an "upstream" party on a "downstream" party and suggest that a pricing mechanism might minimize coordination costs between such parties by penalizing the upstream party for its externalities. Applied to the current context, design will tend to ignore inputs from manufacturing if it pays no penalty for designing a product that is difficult to produce.

Victor and Blackburn (1987) developed an index of concordance that indicates the extent to which the actions of two parties have a positive or negative impact on the other. Differences in rewards and goal priorities among the parties influence the degree to which the parties are willing to cooperate to produce positive outcomes or minimize negative outcomes for each other.

Coordination costs will be lower when the parties are willing to cooperate because they are more likely to settle issues themselves rather than rely on the organizational hierarchy to resolve them.

Differentiation

Coordination of reciprocal interdependence becomes more difficult with increasing differentiation between functions. The difficulty arises because members of the functions cannot understand each other very well or because they disagree on issues that jointly impact them. As Lawrence and Lorsch (1967) indicate, organizational functions can differ on time horizons, technologies, rewards, and interpersonal styles. Such differentiation occurs naturally because of differences in the tasks that functions perform or differences in the backgrounds of people who enter the function.

Approximately 30 managers who attended one of Penn State's Advanced Manufacturing Forum meetings in 1984 were asked to role-play being either a design or manufacturing engineer. Members of each subgroup were asked to describe how they saw their counterparts. Figure 12.2 shows the list of differences that was derived from their descriptions. One can see from this list why personnel from these two functions might have difficulty understanding or agreeing with each other.

PRODUCT ENGINEERS	MANUFACTURING ENGINEERS
Press for tighter tolerances	Press for looser tolerances
Emphasize material costs	Emphasize labor costs
Looser time constraints	Tighter time constraints
Push for state-of-the-art	Pull back to the possible
Materials are integral to design	Willing to accept material substitutes
Jobs vulnerable to business cutbacks	Jobs more secure
Career path steeper (e.g. to VP Eng.)	Career path flatter
Higher salary than ME's	Lower salary than PE's
Rewarded by number of patents	Rewarded by making schedule
Performance criteria fuzzy	Performance criteria clear-cut
Creative orientation	Pragmatic orientation

Figure 12.2. Summary of differences between product engineers and manufacturing engineers.

Differences in education levels between design and manufacturing personnel may hamper communication between them. In our study, 89% of design personnel had bachelor's degrees compared with 59% for manufacturing engineers. Thirty-three percent of design personnel had postgraduate degrees compared with 14% for manufacturing personnel. Differences in professional training and experience also may lead these personnel to approach problems differently. Langowitz (1986) reported on a company where manufacturing engineers could not describe in abstract terms how to design the product for manufacturability; they preferred to see prototypes and then give concrete suggestions. The design engineers, on the other hand, tended to be contemplative before drawing or prototyping. These differences were complicated by the fact that the manufacturing personnel were Japanese and the design personnel were American. The positive side of differentiation is that creative solutions can emerge from a dialogue between design and manufacturing. The negative side is that differentiation can inhibit or complicate communication between them.

Daft and Lengel (1984) suggest that coordination under conditions of high interdependence and high differentiation requires frequent communication between the parties and the use of rich media for exchange of information. Media vary in richness starting from computer printouts and memos to telephones and face-to-face meetings. The most appropriate medium for communicating under high interdependence and high differentiation is the small group. Organizational policies and practices that can influence what medium is best and when will be discussed later in this chapter.

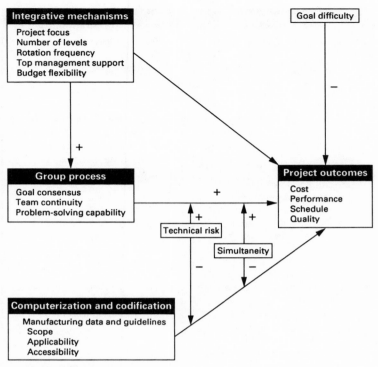

Figure 12.3. The DFM effectiveness model.

A MODEL FOR PREDICTING DESIGN FOR MANUFACTURABILITY EFFECTIVENESS

The policies and practices that companies undertake to improve how design and manufacturing work together can be grouped into three categories: integrative mechanisms (IM), group process (GP), and codification and computerization (CC) of manufacturing data. Integrative mechanisms serve mainly to reduce differentiation, while group process and codification/computerization are alternative methods of achieving the coordination necessitated by interdependence. Integrative mechanisms are likely to be implemented organization-wide, while group process and codification/computerization are more likely to be implemented at the project level; this pattern may vary between companies, however.

A model for predicting DFM effectiveness is presented in Figure 12.3. Essentially, the model proposes that companies will be successful in their DFM efforts to the extent that they have created effective integrative mechanisms, group processes, and codification/computerization of manufacturing data and guidelines. This model is a composite of factors that we have determined—from our interviews, the literature, or both—will lead to successful DFM. Certainly none of the companies in our sample was effective in implementing all or even most of the initiatives we have identified. Depending on the characteristics of the particular product development effort, a moderate level of most of the variables leading to DFM effectiveness, or perhaps a high level on a few of them, may be sufficient. The model also proposes that other factors impact DFM effectiveness either directly or as moderators of one or more of the organizational or project level initiatives.

Integrative Mechanisms

Integrative mechanisms provide a favorable context for projects by reducing differentiation between design and manufacturing. Based on our field experiences and the literature, we have identified a number of such mechanisms, which we have provisionally grouped into six constructs: status parity, project focus, number of levels to a common report, rotation frequency, top management support, and budget flexibility.

Status parity. Manufacturing personnel traditionally have had lower status in organizations than designers (e.g., Ayres, 1984, p. 237). In many organizations, design personnel have seen manufacturing engineers as little more than technicians—people who know the rules about what can and cannot be done but don't understand the science underlying them. The historical basis for this stereotype is that manufacturing engineers have traditionally worked their way up from the manufacturing floor; in some of the companies we studied degreed manufacturing engineers are still in the minority.

The status differences between design and manufacturing engineers are reinforced by differences in pay and office accommodations, and by the ability of designers to ignore or overrule manufacturing engineers in a conflict regarding the manufacturability of a design. Even in companies where manufacturing has formal sign-off authority on designs, their lower status often prevents them from exercising their veto power (Adler, 1989). In one of the companies we studied, manufacturing engineers registered their protests in this situation by signing their names upside down on the approval form.

We propose that such status differences—which serve further to differentiate design and manufacturing—are antithetical to DFM. Companies that have equalized the status of the two functions, in terms of education, pay, accommodations, and influence on design, should be more successful in their DFM efforts. We found in our interviews that most companies were striving to reduce status differences between functions, and a few had virtually eliminated them.

There is considerable support in the literature for the proposition that status differences cause problems in interdepartmental relations. Walton and Dutton (1969) observe that status differences leave the "superior" unit with little incentive to coordinate. McCann and Galbraith (1981) conclude that status asymmetries promote interunit conflict. In an international study of new product development, Bergen and McLaughlin (1988) found that status consciousness was associated with missing project schedules.

Project focus. Virtually all the companies in our study approached new product development with an essentially functional organization structure. It is of course this functional structure that creates the problem of differentiation that DFM efforts try to overcome. In order to coordinate the efforts of those individuals assigned to a given project, most companies augment the functional structure with a matrix or project structure. This involves the assignment of a project manager, the arrangement of weekly meetings, and so on.

In some companies, the project structure is taken very seriously; in others, it is little more than a footnote to a strongly functional structure. As the primary structural integrative mechanism, we propose that a project—as opposed to functional—focus will be associated with more successful DFM efforts.

A project focus is manifested in several ways. First, project managers are given real authority. In some organizations, project managers are little more than meeting-schedulers and report-filers, whereas in others they have the authority to overrule functional managers where the integrity of the project is concerned. A certain amount of authority vested in someone who has the new product's interests at heart will be necessary to achieve DFM goals. Lawrence and Lorsch (1967) note that the project manager's authority should be based on expertise. In a study in the automotive industry, projects with "heavyweight" project managers used fewer engineering hours than those with "lightweight" project managers or none at all (Clark, Chew, and Fujimoto, 1988). A "heavyweight" project manager is likely to have equal or higher pay and seniority than a functional manager, as well as to have budget authority.

Second, the team itself is given a wide range of discretion in matters concerning the new product. It can be quite time-consuming and destructive for a group constantly to be held up while members seek approval for their decisions from their functional organizations. In a study of successful Japanese product development efforts, Imai, Nonaka, and Takeuchi (1985, p. 342) found that "Top management . . . intentionally leaves considerable room for discretion and local autonomy" to the development team. Some of the best-known American DFM success stories—such as the Ford Taurus—also resulted from giving a small team of people broad authority over the design of the product. In an empirical study, Bergen and Mc-Laughlin (1988) found a significant relationship between the autonomy allowed project personnel and performance.

Third, companies with a project focus will provide project-based rewards. That is, a significant portion of an individual's performance evaluation and resulting rewards (pay, promotion, valuable assignments) will depend on the achievement of the project's goals, not narrowly defined functional goals. For example, in some organizations, designers are rewarded for meeting drawing release dates, regardless of whether the designs are manufacturable. In one project we studied, this led to predictable release of drawings that were essentially meaningless owing to their non-producibility. In more successful organizations, both design and manufacturing personnel are held accountable for the success of the project as a whole (Wolff, 1985).

Numerous organization theorists have argued that reward systems materially affect the degree of cooperation among interdependent units (e.g., Walton and Dutton, 1969; McCann and Galbraith, 1981; Brown, 1983). If design and manufacturing are to "take joint responsibility for the outcome of their joint effort" (Clark and Fujimoto, 1989), reward systems must foster regard for the success of the project as a whole. As Imai, Nonaka, and Takeuchi (1985, p. 357) note, "Group evaluation enhances interactions within the group and fosters sharing of skill and knowledge."

Finally, new product development teams in project-focused organizations will be co-located. Our interviewees often noted that working in the same area made it easier to find one another, and the inevitable informal contact led to stronger bonds among team members. In the ideal case, team members would be located in the same suite of offices. In descending order of desirability, they may be located on the same floor, in the same building, at the same site, in the same city, or in the same country. While telecommunications provides methods to ameliorate the problems associated with distance, projects will still be better integrated when members have the frequent opportunity to communicate face-to-face (Ayres, 1984, p. 238).

At a theoretical level, "physical clustering" is seen as an important mechanism for coordination when interdependence is high (McCann and Galbraith, 1981). Imai, Nonaka, and Takeuchi (1985) report that a development team at Fuji-Xerox benefited greatly by the frequent communication their shared office space facilitated. On the other hand, a successful development team at Honda made 2,000 trips over three years—with each 300–mile train trip taking 5 hours—in order to coordinate design and manufacturing. This example fits our impression from our study: distance is a handicap, but it can be overcome.

Number of levels to a common report. While, as noted above, most companies use a primarily functional structure, there is still variation across companies in the extent to which their structures promote or discourage differentiation. One important aspect is how many hierarchical levels exist between project-level personnel in design and manufacturing and the first manager who has authority over both. In some of the companies we studied, there were only one or two levels, in others, five or six. A large number of levels promotes the development of functional kingdoms, where subgoals can be pursued with little information or interference from other functions. Engineers working on new products in such structures will find they have a large chasm to bridge when they attempt to work together to attain DFM goals.

Another reason a large number of levels between project engineers and their first common report is a handicap is that, when issues cannot be resolved by direct cross-functional contact, they are often sent up the hierarchy to the point where a single individual has the authority to decide for both parties (e.g., Galbraith, 1973). The closer this individual is structurally to the location of the conflict, the faster and non-contentiously issues are likely to be resolved. In settings with a large number of functional levels, there is considerable room for delay, politics, and the development of antipathy between the two hierarchies as issues work their way up the hierarchy. Brown (1983) discusses a case where creating common superiors lower in the organization resulted in enhanced coordination between production and maintenance functions.

Frequency of rotation. Companies can to some extent overcome the differentiation inherent in a functional structure by periodically rotating people among the functions. For our purposes, this means an exchange of personnel among design and manufacturing engineering. Companies that undertake such exchanges more frequently will create a climate more conducive to DFM. By spending time in the other function, engineers will develop an appreciation for that function's goals, as well as the constraints under which they work. For example, an engineer who has had to design methods to manufacture new products will be much more aware of the producibility consequences of his or her own designs. In addition, rotated personnel will develop personal relationships with people in the other function, further reducing the differentiation between the two groups.

Rotation of designers into manufacturing (the more common approach) may take any of several forms. Designers may take temporary assignments in manufacturing, and then use their knowledge when they return to design (Thurmond and Kunak, 1988). Or designers may move into manufacturing with the product they have designed (Adler, 1989). Finally, designers may move into manufacturing permanently, as part of a standard career path common to Japanese companies (Westney and Sakakibara, 1986). Each of these patterns reduces functional differentiation in slightly different ways. Miller (1959) proposed that when organization members learn about one another's functions, they are less likely to make unreasonable demands of one another out of ignorance.

While rotation is reportedly common in Japanese companies (Rohlen, 1975), we found it to be unusual in the companies we studied. Although most companies had rotation programs on the books, functional managers working in predominantly "lean" organizations found it hard to justify sparing their people for an educational experience with long-term benefits. In two of the most successful projects we studied, however, design personnel had manufacturing experience, which rendered them much more aware of and sympathetic to manufacturing's concerns.

Top management support. DFM, like any innovative organizational practice, will have a difficult time being successful without support from top management (Thurmond and Kunak, 1988). This support may take a number of forms. Top managers may simply stress the importance of design-manufacturing cooperation in their talks within the firm. They may provide resources for training, travel, or other activities necessary to facilitate DFM. Or they may take a stand in favor of DFM when there is a short-run cost involved, such as postponing a design release until manufacturing's concerns are dealt with. In our study, many managers provided verbal support for DFM; fewer were willing to back it up with resources or difficult decisions. Without tangible support from top management, engineers and middle managers are likely to dismiss DFM as yet another management fad.

Budget flexibility. Our final integrative mechanism is the extent to which design and manufacturing personnel can be assigned to projects in a flexible manner. Inflexibility is particularly problematic in companies working on defense or government contracts. In such settings, manufacturing personnel cannot be budgeted for work on a new product until the design has reached a fairly advanced stage. This prevents manufacturing personnel from having meaningful input on the design in the earliest stages, when it would be most productive, and leads to a reactive posture on the part of manufacturing. Some contracts even encourage this approach by providing funding for redesign only after the product has been released to manufacturing. In general, we expect companies free from such budget constraints to be more successful in their DFM efforts.

Group Process

Many companies pursuing DFM form project teams in order to provide a forum for cross-functional discussion of new product-related issues (Dean and Susman, 1989). Teams typically include design and manufacturing representatives, as well as project managers, quality personnel, and others. What occurs within such teams probably has the most significant direct effect on DFM effectiveness, as they provide the primary coordination method with which design-manufacturing interdependence is addressed. Successful DFM requires not only that such teams be formed, but that they be managed effectively. We have selected three aspects of group process for our model: goal consensus, team continuity, and problem-solving capability.

Goal consensus. There is both conflict and uncertainty inherent in the development of goal structures for new product development teams. The conflict stems from the functional loyalties that members bring to the team (e.g., Ayres, 1984, p. 238). While design is concerned about using advanced materials and creating an elegant design, manufacturing wants to be able to use existing production methods and personnel. The uncertainty is associated with the newness of the product: What performance standards are feasible? How long will it take to design and produce? and so on.

Faced with this challenging scenario, many teams avoid the painful step of trying to attain consensus on the goals of the project. This gives them a weak basis for making decisions, as each participant weighs alternatives against his or her idiosyncratic notion of the team's priorities. We propose that teams that attain goal consensus will be more successful in their DFM efforts. As functional conflict is seldom completely overcome, teams may not be able to attain complete goal consensus. But even teams that attempt to do so will identify their areas of agreement, as well as those areas where they know negotiation will be necessary. The teams in our study exhibited wide variation in goal consensus; some were in virtually complete agreement, while others had almost none.

Team continuity. We propose that DFM efforts will be more successful when the new product team's activities are characterized by a smooth, continuous process. Unfortunately, contextual factors in many organizations conspire to make such a process a rare event. One major enemy of process continuity is turnover within the team. In only a few projects in our study did the team remain intact throughout the project. More often, one or more key members were reassigned in the middle of the project, causing considerable disruption as the replacements got up to speed. Developing both a working knowledge of the product and working relationships with other team members takes a considerable amount of time.

Another impediment to a smooth, continuous process is the tendency of many companies to assign manufacturing engineers to several projects simultaneously, so that they devote only a small proportion of their time to any one of them. Under these circumstances, manufacturing engineers are continually refamiliarizing themselves with the project. They also will be less accessible to other members of the team, as they may often be involved in meetings related to one of their other projects.

Walton and Dutton have noted that "overload . . . may decrease the time available for social interactions that would enable the units to contain their conflict" (1969, p. 74). This seems to be less of a problem with design engineers, who were consistently assigned to fewer projects (often only one) in the companies in our study.

A final aspect of team continuity is the consistency of project team meetings. While projects are often maligned for wasting too much time in meetings, they are the most efficient mechanism for rapidly sharing information and coming to decisions. Project teams that establish and stick to a consistent schedule of meetings will be less likely to have members who work on something irrelevant because they were unaware of changes made in the project. Some of the teams we studied met for a few minutes every day during peak periods, whereas weekly meetings were sufficient for others.

Problem-solving capability. Many authors have portrayed new product and process design as an exercise in information processing, that is, decision-making and problem-solving (Clark and Fujimoto, 1989; Langowitz, 1988; Slusher, Ebert, & Ragsdell, 1989; Dean, Susman, and Porter, 1990). As Clark, Chew, and Fujimoto (1988) put it, "Many [design] objectives cannot be met with obvious alternatives. When this happens, the engineers have a problem, and the problem-solving cycle begins" (p. 10).

Engineers attempting to solve problems in the context of new product development face two obstacles. The first is uncertainty, which is "the difference between the amount of information required to perform the task and the amount of information already possessed" (Galbraith, 1977). The second is equivocality, "the existence of multiple and conflicting interpretations about an organizational situation" (Daft and Lengel, 1986). Uncertainty and equivocality are products of both the novel technology and the need to coordinate across departments differentiated in their perceptions, experiences, and values, and are quite likely to occur when new products are launched (Daft and Lengel, 1986). Thus for DFM to be successful, team members must overcome the challenge of uncertainty and equivocality using collaborative problem-solving practices.

Merely putting a group of people together and calling them a team hardly ensures that they will be able to accomplish this perplexing task: New product teams vary substantially in their ability to interact productively and creatively to solve design problems. In general, an atmosphere of coordination and communication between functions has been linked to less severe mismatches between new products and production capabilities (Langowitz, 1988).

There are a number of specific aspects of the way such groups approach problem-solving that we propose will lead to effective DFM. First, group members must be willing to share provisional information with one another and to react on the basis of such information (Clark and Fujimoto, 1989), as well as to treat their decisions as tentative until late in the project. This will be easier if team members trust one another, both personally and professionally. Mistrust has been empirically associated with restriction and distortion of information (Zand, 1972), which may greatly hamper problem-solving. Sharing tentative information, acting upon it, and considering decisions tentative will render new product teams much more flexible in responding to problems.

Second, group members must strive to see things from one another's functional perspective, not just from their own. Design and manufacturing personnel find this quite difficult, given the degree of differentiation that characterizes most companies. This is why the integrative mechanisms discussed above, which reduce differentiation, are so necessary to provide the foundation for effective team interaction. Third, effective teams should promote team-wide involvement in discussions, which will provide a broader base of solutions to problems. Finally, effective groups will seek integrative solutions, which meet everyone's needs, rather than trying to "win" by serving one groups needs over another's, or compromising so that no one's needs are completely met (e.g., Walton and McKersie, 1965).

There was a wide range in the extent to which teams we studied practiced these effective problem-solving techniques. Some groups were very effective at problem-solving, but others were very poor, which resulted in both bad products and bad feelings within the group. Effective problem-solving within new product teams is facilitated by several integrative mechanisms, including status parity, project focus, and frequent rotation. We must point out that effective problem-solving on DFM-related issues does not take place only in regularly scheduled project meetings. In fact, only 27% of the respondents in our study reported that comments on drawings were typically made in such meetings. Thirty percent of respondents reported typically making comments on drawings in special design review meetings. The largest number of comments on drawings (40%) was made in ad hoc one-on-one meetings.

Relationship between integrative mechanisms and group process. In addition to their direct effect on DFM effectiveness, integrative mechanisms are likely to influence outcomes indirectly through their influence on group process. Project focus is the construct within integrative mechanisms that is most likely to influence group process. In particular, project focus is most likely to influence goal consensus and/or problem-solving capability.

Codification and Computerization

Codification and computerization (CC) refers to the existence, applicability, and accessibility of data and guidelines to assist in achieving manufacturable designs. IM and GP refer to the context and process of DFM, respectively; CC refers to its tools. CC represents an alternative coordination mechanism to group meetings, with different advantages and disadvantages to be discussed below. There are two broad categories within CC, manufacturability data and manufacturability guidelines.

Manufacturability data. As its name implies, this category refers to the codification and computerization of data that will assist in designing producible products. Companies that have an extensive, widely applicable, and easily accessible database concerning manufacturing are likely to be more successful in their DFM efforts than those that do not.

Several dimensions characterize databases more likely to aid DFM performance substantially. The first of these is scope, which is the number of manufacturing topics about which the database includes information. While some databases may include only machine tolerance data, others may also include information about materials, quality tests, costs associated with various processes, and so on. Clearly those that cover a wider scope of topics will be more helpful to design personnel in their DFM efforts. In our study, the most common kind of data available pertained to quality.

A second dimension differentiating manufacturability databases is their range of applicability, that is, the number of different types of products to which they are applicable. Sometimes in our interviews we were told about the existence of a certain database, only to be told later that it was only applicable to one type of product. In fact, given the recentness of efforts in this area, most manufacturability databases probably fit this description. Broadly applicable databases will be more effective in improving DFM.

A third dimension relevant to manufacturability databases is their accessibility. While some databases can be accessed with only the touch of a few keys on the designer's terminal, others require searching through stacks of paper records in company archives. Needless to say, the former will have a great deal more impact. But computerization is not always the answer: a database with poor documentation that is not accessible from the designer's desk will be less helpful than a simple hard copy of data that is within reach. The key issue is how long and how much effort it takes the designer to access the information he or she needs.

Manufacturability guidelines. Manufacturability guidelines are in a sense data once removed. Many companies have approached the DFM problem by adding manufacturability rules into their existing design guidelines or standards (Thurmond and Kunak, 1988; Dean and Susman, 1989). Some companies use generic guidelines from an external source, such as commercial assemblability packages; others devise their own guidelines from their unique production experiences. To the extent that the plant's capabilities are in any way unusual, the latter will be more beneficial.

The function of manufacturability guidelines is much the same as that of data: to provide reliable guidance to design engineers trying to achieve DFM goals. Thus it makes sense that the same variables should differentiate better and worse sets of guidelines and databases. That is, sets of guidelines are likely to be more effective to the extent that they are broad in scope, i.e., deal with a wide range of DFM concerns; widely applicable, i.e., are relevant to many types of products, and easily accessible, i.e., can be reviewed with little expenditure of time or effort on the part of designers. Guideline accessibility can be promoted by computerization, especially when they are integrated into designers' CAD/CAE workstation software (Adler, 1989).

The obvious advantage of guidelines and data (at least in the short term) is that they require little interaction between functions. But a disadvantage of manufacturability guidelines is that, in the context of status disparity, design engineers may not take them seriously. Argyris (1964) and Dalton (1959) argued that conflict results when a lower-status unit sets standards for one of higher status. In one of our earliest investigations of DFM, we found that designers were reluctant to follow manufacturability guidelines, which they suspected were arbitrary and unnecessary (Susman and Dean, 1986).

Goal Difficulty

Goal difficulty is defined as the size of the difference in cost, quality, lead time, or performance between what the company targets for the product and what it achieved most recently on similar products. For a variety of reasons, some projects set more difficult goals than did other projects, particularly regarding lead time and cost. Eighteen of the twenty-one projects were trying to improve at least one of these four goals substantially beyond what they had achieved in the past. For example, some projects were trying to reduce their schedule by 50% or more or improve their quality by 200% or 300%. Eleven of the projects were seeking improvement of similar magnitude in two to four of these goals at once. Mansfield et al. (1971) had shown that trade-offs normally exist for such goals, at least under existing organizational arrangements. DFM is an attempt to modify such arrangements so that such trade-offs might be transcended.

If our model were to ignore such differences, and equate all projects in terms of goals set, it would be biased in favor of projects with easy goals. Therefore, goal difficulty is included in the model as a main effect with a negative relationship to project outcomes. That is, all things being equal, companies with difficult goals will be less likely to achieve them.

Moderating Conditions

So far the model we have presented has been relatively straightforward, insofar as higher levels of the variables in our three categories are expected to enhance the chances for success of DFM projects. There has been so far no recognition of any differences among projects that would moderate the effects we have discussed. As Slusher, Ebert, and Ragsdell (1989, p. 3) note, "There is some agreement that design projects should be managed in ways contingent on the special characteristics of the particular project." In this section, we discuss two such characteristics that make our model both more realistic and more complex.

Technical risk. New products differ substantially in their degree of technical risk, defined as the extent to which they rely on advances in state of the art technology (Cochran, Patz, and Rowe, 1978; Thurmond and Kunak, 1988; Slusher, Ebert, and Ragsdell, 1989; Clark, Chew, and Fujimoto, 1988). Some projects attempt only modest advances, or none at all, while others rely on technology whose feasibility is highly questionable (e.g., Graham, 1986). Since DFM is concerned with the match between product and process technology (Langowitz, 1988), advances in either will create technical risk. It has been widely observed (e.g., Langowitz, 1988) that greater technical risk presents formidable problems for project teams. Why?

From an information-processing perspective, projects high in technical risk may be seen as presenting higher degrees of uncertainty and equivocality to project personnel (Cochran et al., 1978). Technology has long been seen as a source of uncertainty (McCann and Galbraith, 1981), and Daft and Lengel (1986) contend that both uncertainty and equivocality are particularly likely to occur "during times of rapid technological development" (p. 558).

Once technical risk's implications for uncertainty and equivocality are understood, its theoretical link to our model is straightforward. When uncertainty and equivocality are high, a great deal of information processing must take place in order to address them. How can this be accomplished? According to McCann and Galbraith (1981), "Rules and programs become less effective in more uncertain situations. [There is] a greater need for mutual adjustment or coordination if response times are to be reduced and conflict managed" (p. 70). Similarly, Daft and Lengel (1986) write: "Information systems do not reduce equivocality because equivocal issues are not easily measured and communicated through impersonal systems. . . . For non-routine technology [involving uncertainty and equivocality], group meetings will be a primary source of information processing" (pp. 562–564).

Thus projects marked by a high degree of technical risk will find that their group processes, which are well-suited to reducing uncertainty and equivocality, are more critical than in projects with relatively little risk. Conversely, manufacturability guidelines and information systems will be commensurably less critical. We have included technical risk in the model as a moderator of the relationships of both GP and CC with project outcomes. We propose that there will be a stronger relationship between group process and DFM effectiveness in high-risk projects, and a weaker relationship between CC and project outcomes (see Figure 3). This addition to the model squares nicely with Langowitz's (1988) conclusion, "Particularly on highly technically ambitious projects . . . emphasis should be placed on including and working with manufacturing in the development process" (p. 52).

Simultaneity. As we noted at the beginning of this chapter, very ambitious DFM efforts often feature "simultaneous" or "concurrent" engineering, in which the design of the manufacturing process is begun well before product design is complete (Imai, Nonaka, and Takeuchi, 1985; Clark and Fujimoto, 1989). It is well known that overlapping stages of product development can lead to substantial reductions in lead time (Mansfield et al., 1971). But it is also being discovered that managing such "overlapping" processes is a challenging task: "the burden of managing the process increases exponentially" (Imai, Nonaka, and Takeuchi, 1985, p. 350). Clark and his colleagues have consistently reported that simultaneous engineering requires a particularly intense style of information processing, without which simultaneity "may cause serious side effects, such as lower design quality, unintended schedule delay, and loss of motivation" (Clark and Fujimoto, 1989). How can this observation be connected theoretically with our model?

Simultaneity can be understood theoretically as increasing interdependence beyond the normal level that characterizes the design-manufacturing interface. This occurs in two ways. First, what was traditionally a relationship of sequential interdependence (Thompson, 1967) is now reciprocal, as design depends on manufacturing for frequent updates on how its plans will influence product design. Second, interdependence is increased when the time it takes work to move back and forth between units decreases (Gerwin and Christoffel, 1974). This change clearly characterizes simultaneous engineering, which has often been compared to a rugby game, in which players run together downfield, constantly passing the ball back and forth among them (e.g., Imai, Nonaka, and Takeuchi, 1985).

With the link between simultaneity and interdependence in place, we can rely on the theoretical literature on interdependence for help in determining the implications of simultaneity for our model. After reviewing studies such as those of Lawrence and Lorsch (1967) and Van de Ven, Delbecq, and Koening (1976), McCann and Galbraith (1981) concluded: "All of the studies suggest that portfolios of less formal, more coordination-oriented, and individual and group-level strategies gain emphasis as interdependence increases" (p. 78). Similarly, Victor and Blackburn (1987) concluded that "increases in interdependence should be associated with the selection and effectiveness of more lateral, organic, coordination strategies" (p. 494).

In terms of our model, this means that as simultaneity increases, group processes become more crucial for effective DFM, and information systems and guidelines (more formal and rule-oriented) become less crucial, perhaps even counterproductive (Adler, Chapter 9; Slusher and Ebert, Chapter 8). Thus our model (see Figure 12.3) proposes that simultaneity should have a positive moderating effect on the relationship between group process and project outcomes, and a negative moderating effect on the relationship between codification/computerization and project outcomes. This addition to the model is quite consistent with Clark and Fujimoto's observations from their fieldwork: "For effective operation of overlapping, problem-solving must be integrated through dense, reciprocal flows of information" (p. 8).

LIMITATIONS OF THE MODEL

While we have struggled in conceptualizing our model to be as comprehensive as possible and to be faithful to both our field observations and the literature, it has a number of limitations that future empirical and conceptual efforts will need to overcome. First, it ignores the role of suppliers—which often provide the majority of parts for a project—in the DFM process. For any firm that relies heavily on purchased parts, comprehensiveness would require a more complex model of interorganizational problem-solving.

Second, the model overlooks the long-term implications for manufacturing of being "stretched" by design demands (Langowitz, 1988). While in the short term projects will be successful if design's demands fall well within manufacturing's current capabilities, in the long run manufacturing will be better off if they are forced to pull themselves ahead to meet design challenges. Looked at from another perspective, this implies that, if DFM is to be effective in the long run, design should not do all the compromising.

Third, we have ignored the influence of individual differences, which numerous authors have linked to variation in the success of interdepartmental coordination (e.g., McCann and Galbraith, 1981; Brown, 1983). Certainly differences in interpersonal skills and attitudes will influence DFM effectiveness. Our model does not capture these differences, except to the extent that they result in enhanced problem-solving capacity.

Finally, the model has yet to be empirically tested. Since it was developed on the basis of our 12–company study, it would be inappropriate to test the model on this same sample (even if it were big enough, which it isn't). A self-administered questionnaire has been developed on the basis of the structured interviews we conducted in these companies. It has been mailed to design and manufacturing engineers in several hundred manufacturing firms. The large database that responses to these questionnaires will generate will permit us to take the next step toward understanding the factors that contribute to design for manufacturability effectiveness.

NOTE

1. Although twenty-four projects (two projects per company) were initially selected for the study, the final number of projects from which data were collected was twenty-one. One company was not able to participate in the second site visits due to internal matters that were unrelated to the study. Another company decided not to have interviews conducted for one of the two projects it had initially selected because it thought the project was less relevant for generalization about its DFM activities.

REFERENCES

Adler, P. S. "Interdepartmental Interdependence and Coordination: The Case of the Design/Manufacturing Interface." Department of Industrial Engineering and Engineering Management, Stanford University, 1989.

Argyris, C. Integrating the Individual and the Organization. New York: Wiley, 1964.

Ayres, R. U. The Next Industrial Revolution; Reviving Industry Through Innovation. Cambridge, Mass.: Ballinger, 1984.

Bergen, S. A., and C. P. McLaughlin. "The R&D/Production Interface: A Four-Country Comparison." International Journal of Operations and Production Management 8(7), 1988, 5–13.

Boothroyd, G., and P. Dewhurst. "Product Design for Manufacture and Assembly." Manufacturing Engineering, April 1988, 42–46.

Brown, L. D. Managing Conflict at Organizational Interfaces. Reading, Mass.: Addison-Wesley, 1983.

Clark, K. B., W. B. Chew, and T. Fujimoto. "Product Development in the World Auto Industry." Brookings Papers on Economic Activity 3, 1988, 729–771.

Clark, K. B., and T. Fujimoto. "Overlapping Problem-solving in Product Development." In Managing International Manufacturing, ed. K. Ferdows. North Holland: Elsevier Science Publishers B.V., 1989, 127–152.

Cochran, E. B., A. L. Patz, and A. J. Rowe. "Concurrency and Disruption in New Product Innovation." California Management Review 21(1), 1978, 21–34.

Daft, R. L., and R. H. Lengel. "Information Richness: A New Approach to Managerial Behavior and Organizational Design." Research in Organizational Behavior 6, 1984, 191–233.

Dalton, M. Men Who Manage. New York: Wiley, 1959.

Dean, J. W., Jr., and G. I. Susman. "Organizing for Manufacturable Design." Harvard Business Review 67(1), January/February 1989, 28–36.

Dean, J. W., Jr., G. I. Susman, and P. S. Porter. "Technical, Economic, and Political Factors in Advanced Manufacturing Technology Implementation." Journal of Engineering and Technology Management 7, 1990, 129–144.

Galbraith, J. R. Designing Complex Organizations. Reading, Mass.: Addison-Wesley, 1973.

Galbraith, J. R. Organizational Design. Reading, Mass.: Addison-Wesley, 1977.

Garvin, D. A. "What Does 'Product Quality' Really Mean?" Sloan Management Review, Fall 1984, 25–43.

Gerwin, D., and W. Christoffel. "Organizational Structure and Technology: A Computer Model Approach." Management Science 20, 1974, 1531–1542.

Graham, M. B. W. RCA and the Videodisc: The Business of Research. Cambridge, England: Cambridge University Press, 1986.

Hauser, J. R., and D. Clausing. "The House of Quality." Harvard Business Review, May-June 1988, 63–73.

Imai, K., I. Nonaka, and H. Takeuchi. "Managing the New Product Development Process: How Japanese Companies Learn and Unlearn." In The Uneasy Alliance, eds. K. B. Clark, R. H. Hayes, and C. Lorenz. Cambridge, Mass.: Harvard University Press, 1985, 337–381.

Langowitz, N. S. "An Exploration of Production Problems in the Initial Commercial Manufacture of Products." Research Policy 17, 1988, 43–54.

Langowitz, N. S. Plus Development Corporation (A). Harvard Business School, (9–687–001), 1986.

Lawrence, P. R., and J. W. Lorsch. Organization and Environment. Boston: Division of Research, Graduate School of Business Administration, Harvard University, 1967.

Mansfield, E., J. Rapoport, J. Schnee, and S. Wagner. Research and Innovation in the Modern Corporation. New York: Norton, 1971.

McCann, J. E., and J. R. Galbraith. "Interdepartmental Relations." In Handbook of Organiza-

tional Design, eds. P. C. Nystrom and W. H. Starbuck. New York: Oxford University Press, 1981, Vol. 2, 60–84.

Miller, E. J. "Technology, Territory, and Time." *Human Relations* 12, 1959, 243–272.

Qualls, W., R. W. Olshavsky, and R. E. Michaels. "Shortening of the PLC-An Empirical Test." *Journal of Marketing* 45, Fall 1981, 76–80.

Reinertsen, D. G. *Who Dunnit? The Search for the New Product Killers*. New York: McKinsey and Co., 1983.

Rohlen, T. P. "The Company Work Group." In *Modern Japanese Organization and Decision-Making*, ed. E.F. Vogel. Berkeley: University of California Press, 1975, 185–209.

Slusher, E. A., R. J. Ebert, and K. M. Ragsdell. "Contingency Management of Engineering Design." Presented at the International Conference of Engineering Design, Harrogate International Centre, United Kingdom, 1989.

Stalk, G., and T. M. Hout. *Competing Against Time: How Time-Based Competition is Reshaping Global Markets*. New York: The Free Press, 1990.

Susman, G. I., and J. W. Dean, Jr. "A Case Study of Four Projects Supported by the Industrial Modernization Incentives Program of the Department of Defense." University Park, PA: Center for the Management of Technological and Organizational Change Working Paper Series, No. 86–10, 1986.

Susman, G. I., and J. W. Dean Jr. "Strategic Uses of CIM in the Emerging Competitive Environment." *Computer Integrated Manufacturing*, 1989, 133–138.

Thompson, J. D. *Organizations in Action*. New York: McGraw-Hill, 1967.

Thurmond, R. C., and D. V. Kunak. "Assessing the Development/Production Transition." *IEEE Transactions on Engineering Management* 35(4), 1988, 232–238.

Van de Ven, A. H., A. L. Delbecq, and R. Koenig, Jr. "Determinants of Coordination within Organizations." *American Sociological Review* 41, 1976, 322–338.

Victor, B., and R. S. Blackburn. "Interdependence: An Alternative Conceptualization." *Academy of Management Review* 12(3), 1987, 486–498.

Walton, R. E., and R. B. McKersie. *A Behavioral Theory of Labor Negotiations*. New York: McGraw-Hill, 1965.

Walton, R. E., and J. M. Dutton. "The Management of Interdepartmental Conflict: A Model and Review." *Administrative Science Quarterly* 14, 1969, 73–84.

Westney, D. E., and K. Sakakibara. "Designing the Designers; Computer R&D in the United States and Japan." *Technology Review*, April 1986, 24–69.

Wolff, M. F. "Bridging the R&D Interface with Manufacturing." *Research Management*, January-February 1985, 9–11.

Zand, D. E. "Trust and Managerial Problem-solving." *Administrative Science Quarterly* 17, 1972, 229–239.

13

ORGANIZATIONAL CONTEXT BARRIERS TO DFM

JEFFREY K. LIKER and MITCHELL FLEISCHER

Sue, a product engineer at XYZ Motors, sits at a CAD system and develops a design for a piston rod. She develops a preliminary design, analyzes it, has prototypes made, tests the prototypes, refines the design, develops new prototypes, tests them, and finally asks manufacturing if they can produce the part. Manufacturing complains that the tolerances are too tight and that certain shapes are difficult to machine. Sue makes some of the changes, but vetoes others that might interfere with product functioning, and releases the design to manufacturing. Manufacturing discovers many more problems in early production runs and, through a formal engineering change system, requests changes that must be approved by product engineering. Sue agrees to some and vetoes others. Manufacturing secretly makes some of the vetoed changes anyway. Thus, the battle lines between design and manufacturing are drawn and the enemy engaged. . .

This scenario exemplifies the traditional design process in which the design is thrown back and forth over the ''brick wall'' between design and production. The result of this process is extended design lead times, high reject rates, extensive paperwork, high warranty costs, and antagonism between design and manufacturing. One method to prevent this is concurrent engineering, in which product and process design overlap, rather than taking place serially (Clark and Fujimoto, 1989b). A common approach to attempting this overlapping problem-solving is to get the product engineer to meet frequently with the manufacturing engineer early in the design process. While this seems simple enough, in reality the design of a complex product such as a car or even the engine of a car involves thousands of design and manufacturing personnel working within different departments, company divisions, and outside firms (e.g., original equipment manufacturers (OEM), vendor and supplier firms); operating under different value and reward systems; speaking different technical languages; and each pursuing a different agenda.

This chapter will illustrate some of the organizational complexities that create barriers to enacting successfully simple DFM prescriptions by describing how two divisions of a large U.S. automotive manufacturer sought to overcome these barriers. We describe the structural, political, and cultural context of product-process design in this company, that has been actively seeking to encourage DFM. In so doing, we

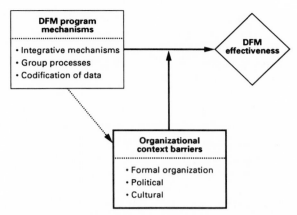

Figure 13.1. Organizational context barriers to DFM.

will attempt to move beyond a static picture of organizational context and begin to develop a dynamic view of organizational process.

Many of the chapters in this book touch on organizational context issues and hence there will necessarily be some overlap between this chapter and those others. For example, Susman and Dean (Chapter 12) distinguish between group processes, integrative mechanisms, and codification of information and argue that all three must work together for successful DFM. Our "organizational context" is not unlike their concept of "integrative mechanisms." However, while "integrative mechanisms" focuses on specific, implementable organization policies and practices, we are interested in the less easy to manipulate unit goals, values, language, and symbols that make up the organization's culture and politics.[1] While earlier chapters focused on tools, methods, and organizational arrangements used to foster DFM, this chapter will help explain some of the difficulties involved in implementing those tools and methods. In other words, we will get below the surface of specific policies and practices and look at the political and cultural dynamics that are sometimes referred to as the "informal organization." We will also look at the extent to which DFM programs are able to penetrate these underlying dynamics and make a real change in the informal organization. In this chapter we address two questions:

1. What was it about the prior organizational context that forced industry to create special programs aimed at DFM?
2. To what degree are DFM programs actually changing the organizational context to successfully support DFM?

The distinction between what we call "DFM mechanisms," which includes Susman and Dean's integrative mechanisms, group processes, and codification of data, and the organizational context is depicted in the model in Figure 13.1. This model portrays "organizational context" as a set of contingencies that influence the success of management's DFM programs. That is, the success of DFM will depend on the supportiveness of the broader organizational context. The model also suggests that DFM mechanisms can help reshape the organizational context to make it more supportive of efforts to design products with manufacturability in mind. We use a dotted arrow, as it remains to be seen whether DFM mechanisms can actually impact the organizational context.

The working hypothesis going into our field research was that the organizational context of design and manufacturing in many large U.S. corporations presents serious barriers to DFM. The contexts we have observed have been primarily traditional, bureaucratic organizations. Within this context, organizational barriers include structural complexity of the formal organization, divergent political interests of design and manufacturing, and incongruent cultural values and symbols (see Figure 13.1). By organizational structure complexity we mean the numbers, geographical separation, organizational boundaries, and structural differentiation of personnel involved in product and process design and manufacturing.

The cases for this chapter are based largely on interviews with 17 people involved in DFM programs, from both the product design and manufacturing side, in two divisions of an automotive manufacturer referred to here as AUTOCO. The respondents selected were all directly involved in a DFM "experiment" and most had worked on at least one major design project before the company instituted its DFM programs. Thus, they were able to contrast the new with the old. The length of the interviews ranged from one to three and one-half hours. We developed an interview schedule touching on the issues suggested by Figure 13.1 but allowed ourselves the flexibility to explore potentially interesting topics that arose in the interview situation. The general flow of the interviews was as follows: Where were the barriers to DFM in past, traditionally run projects? Where are you today in DFM? What were the impacts of the changes made? What still remains to be done? If respondents answered our questions in generalizations we prompted for detail and examples.

In addition to these focused case studies, we draw on experience and research with the design staffs of more than 20 other companies (Fleischer and Liker, 1990; Liker, 1988; Liker and Hancock, 1986). This past research focused on the design process, though not on DFM per se. However, the knowledge gained was still useful for interpreting the observations in the two case examples of DFM. From this past work we believe that most of the experiences of AUTOCO are not unique, but rather are typical for large companies with complex products.

The organization of the chapter is as follows: First, we present basic organizational theory that provides a context for our understanding of DFM. Second, we provide background information about the case examples of DFM in the two automotive plants. Third, we present an overview of the past organizational contexts and design processes in these organizations before there was any targeted effort to encourage DFM. Fourth, we discuss the new management initiatives designed to break down these old barriers in the two plants—their accomplishments and the barriers that remain even with major organizational changes. Finally, we step back from the specific cases and evaluate what was learned about organizational contexts that inhibit and support effective DFM.

ORGANIZING FRAMEWORK

For this analysis we began with an open-systems perspective (Nadler and Tushman, 1980; Katz and Kahn, 1966). The design process transforms inputs into outputs with the ultimate goal of satisfying customer needs. The open-systems model emphasizes interdependence—both interdependence of the components of the transformation process (in this case the design process) as well as interdependence with the external environment.

Internally, open-systems theory suggests we should be seeking congruence between formal and informal organizational arrangements. For example, unless there is a "fit" between the reward structure and the behaviors required of product and process engineers for a particular DFM program, the program is not likely to be effective in meeting its goals or to last very long.

Externally, the best approach to DFM will depend on both the nature of customer demand, the output side, and the process of selecting inputs. On the output side, if customer demand is stable and relatively insensitive to quality (as was true to a large degree in the postwar period in the U.S.) then a mechanistic, bureaucratic organization that emphasizes efficiency can be very effective (Burns and Stalker, 1961; Lawrence and Lorsch, 1967). By contrast, DFM has come to the fore as a management priority in a time of intense competition, increasing consumer demand for high quality products, and strong pressure for rapid response to changing consumer tastes. This suggests a need for more adaptive organizational forms.

On the input side, key inputs to the design process include suppliers and personnel. Suppliers include companies that produce manufacturing equipment (referred to here as equipment vendors) and those that manufacture component parts that go into the final assembled product (referred to as parts suppliers). Thus, when we speak of "manufacturing," we must include the myriad of outside suppliers who manufacture much of the product and those vendors who design and build much of the manufacturing equipment. The organizational arrangements and politics of supplier-customer relations therefore are key factors in the organizational context of DFM. Another key environmental input is personnel. As we will see, the recruitment process for manufacturing engineers and product engineers can be quite different; and these people enter their respective roles with different, and often conflicting, expectations, career orientations, and values—important factors in the success of a DFM effort that requires cooperation between design and manufacturing.

Our model in Figure 13.1 emphasizes organization structure, political systems, and cultural systems, as factors conditioning the effectiveness of DFM policies and practices. An exhaustive review of the literature in these three areas is beyond the scope of this chapter, but a brief discussion of how we will use these concepts follows.

Formal Organization Barriers

Despite changes in the environment, companies in the U.S. have been slow to change their internal design practices. The organization of design is highly differentiated with an extensive division of labor and a linear sequential design process. Burns and Stalker (1961) would characterize much of what we have seen in the organization of product design as relatively "mechanistic" organizations governed by many levels of authority, strong hierarchies and weak horizontal communication, and extensive rules and standard operating procedures. In contrast to the more flexible "organic" organization, which works well in a changing environment requiring innovative behavior, mechanistic organizations are strong in stable environments when routine tasks need to be performed efficiently.

While it is easy to point fingers at top management, in fact they have only limited control over the beast that was created through decades of bureaucratization. Max Weber in the late 1800s accurately forecast the growth and dominance of the bureaucratic form of organization because of its great efficiency. But it was not until after World War II that sociologists began to alert us to the unintended consequences of bureaucracy (Merton, 1968; Selznick, 1949; March and Simon, 1958). These unintended consequences take the form of self-defeating feedback loops that implant in bureaucracies the seeds of their own destruction.

For Merton (1968: chap. 8) the main feedback loop concerns the way bureaucrats respond to their customers. Bureaucratic organizations attempt to control and coordinate activities of different parts of the organization by developing extensive rules and standard operating procedures (SOPs). The SOPs are designed for "average" circumstances. Naturally, real circumstances often differ from average. Service providers under bureaucracy do not have the authority to violate SOPs without going through the chain of command to get higher approval. Since this is cumbersome, and in some cases almost impossible, service providers often refuse to act flexibly even when their customers give reasonable rationales for altering the rules. As a defensive reaction, service providers tend to blame the customers for not doing their job to fit within the rules. This leads to conflict between the servers and the very people they are supposedly serving. Service providers begin to stereotype the customers (e.g., lazy, selfish, shortsighted, and so on). Customers perceive the bureaucrat as arrogant and haughty. Thus, the self-reinforcing loop causes conflict between the service provider and those served. The more circumstances differ from average, the more conflict will result and the more unresponsive the system.

Figure 13.2 illustrates the distinction between the design approaches resulting from the bureaucratic and the open-systems models. The bureaucratic model leads to a design process organized as a set of separate specialties, each shown as a "chimney." The chimney metaphor, used by AUTOCO to critique its own past practices, suggests a separation of functions and an upward orientation of the functions. For example, the service provider in Merton's model is in one chimney and is more interested in satisfying his or her boss higher up in the chimney than serving the needs of the customer in the next chimney. Satisfying the boss may involve sending bogus statistics, or smoke, up the chimney to provide the appearance of effective performance despite the reality of frustrated customers.

The open-systems model suggests the need for a customer focus by everyone in the organization, a perspective shared by virtually all contemporary quality approaches (Deming, 1982; Juran, 1989; Schonberger, 1990; Ishikawa, 1985). The term customer in this context does not refer only to the consumer who buys the car, but to all stakeholders who depend on the employee's services. For example, the product engineer should be focused on providing quality service to the process engineer. In companies operating under the bureaucratic model the product engineer begins to take on the characteristics of Merton's bureaucrat and establishes a position that is antagonistic to manufacturing. The goal of most organizational approaches to DFM is to break down the chimney structure and shift to a customer focus.

The difficulty of making this transformation is that the chimney structures have

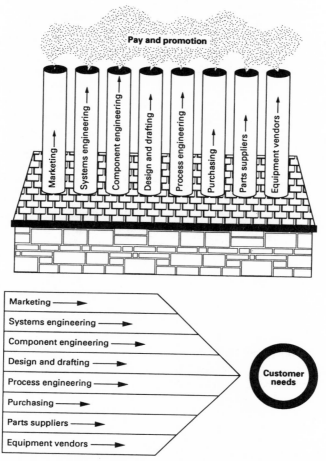

Figure 13.2. Chimney model: upward orientation within functions. Customer-focus model: internal & external customer orientation.

been established and reinforced over decades. A change in the structure challenges cultural and political systems that have emerged and gained strength in this time. Selznick's (1949) model provides an illustration of the dynamics by which subunits proliferate and develop their own bases of power. This is illustrated below as we discuss the conflicting goals among internal product engineers, process/manufacturing engineering, purchasing, and outside parts suppliers and equipment vendors.

Political Barriers

The design process involves information transformations to make a series of decisions. When Max Weber described the emerging organizational form of bureaucracy, he envisioned a system of policies and procedures in which decisions were made rationally according to clear and calculable rules and systems of logic (Weber, 1946). In bureaucracies, specialist knowledge is isolated in specific departments, and specialists are selected according to formal credentials. Thus, decisions are made rationally by specially trained and credentialed individuals who use rational and logical means to make the decision. Indeed, Weber lamented the loss of the stuff of human life such as emotions and cultural symbols as we moved toward a machinelike existence.

Many scholars over the years have demonstrated the limitations of this overly rational view of organizations (e.g., March and Simon, 1958). For example, the rational view assumes a consistent set of goals against which decision-makers can optimize, an assumption that rarely fits organizational life. For example, Allison (1971) analyzed the decisions made about the Cuban missile crisis and argued that much of what transpired cannot be explained using a rational-choice model; the decisions made can better be understood by considering the often conflicting interests of political actors within each of the countries involved.

Politics and business decision-making are often thought to be incompatible. Indeed, Mintzberg (1983, p. 172) defines political behavior as outside legitimate systems of influence: "politics refers to individual or group behavior that is informal, ostensibly parochial, typically divisive, and above all, in the technical sense, illegitimate—sanctioned neither by formal authority, accepted ideology, nor certified expertise."

Pfeffer (1981) provides a different view of politics. He argues that under certain conditions the use of political mechanisms by organizational members to influence organizational decisions is the only way to make the decision. The conditions below, taken together, support the use of power and politics as a means of making decisions. If any one of the conditions is missing, the chances that politics will come into play in decision-making is reduced considerably.

1. Important decision issues
2. Interdependence
3. Conflicting goals
4. Scarcity of resources
5. Diffuse distributions of power

We learned in our case studies that design decision-making has each of the five properties above. Design decisions have important implications for many actors in the system. As discussed above using the chimney metaphor (Figure 13.2), these actors are organized into different departments responsible for different but interdependent phases of the design and manufacturing process. Pfeffer (1981) points to organizational differentiation as one of the major sources of conflicting goals, and we will illustrate the conflicting goals of different functional units in the design process below. Resources are scarce in any design program and different organizational units involved in the design process are vying for what they perceive as a limited quantity of organizational rewards in the form of budget, staffing, equipment, and career opportunities. Finally, the distribution of power in making detailed design decisions is typically diffuse, rather than centralized in a single hierarchy.

Career interests and organizational politics. Morgan (1986) views political interests as a confluence of task demands, career aspirations, and extramural life-style concerns. He notes that complete convergence of these three sets of interests is rare and argues that this is a central reason organizational rationality in the pure form is virtually nonexistent. As we will discuss below, our case studies revealed that the system of career advancement in design and manufacturing was a major inhibitor to the goals of design for manufacturability.

Manufacturer-supplier relations and organizational politics. Emerson (1962) emphasizes the central role of dependence in power relations. He writes (p. 32) that ". . . the power to control or influence the other resides in control over the things he values, which may range all the way from oil resources to ego support. In short, power resides implicitly in the other's dependence." Pfeffer and Salancik (1978) have carried this notion of power dependence further to the study of interorganizational relationships. The case of the relationship between suppliers and the manufacturers who are their customers can be viewed as a power-dependence relationship. Suppliers are dependent on manufacturers for their revenue, the lifeblood of their existence. Kamath and Liker (1990) demonstrated that automotive suppliers who are highly dependent, as measured by the percentage of their business that is automotive related, are particularly responsive to the needs of their customer for product innovation. The case studies described here reveal the central importance of suppliers in product-process design and suggest that some common purchasing practices have created major barriers to DFM.

Cultural Barriers

According to Schein (1984, p. 375), culture is "the pattern of basic assumptions that a given group has invented, discovered, or developed in learning to cope with its problems of external adaptation and internal integration, and that has worked well enough to be considered valid, and therefore, to be taught to new members as the correct way to perceive, think, and feel in relation to those problems."

Organizational culture can be analyzed at three different levels (Schein, 1984). Visible artifacts, which include architecture, office layout, manner of dress, visible behavior patterns, and public documents (such as charters), are the easiest to observe, though they provide only clues as to the underlying reasons why a group behaves as it does. At a deeper level are values that cannot be observed directly but can often be inferred from interviews with organizational members. However, what we hear often reflects "espoused values" rather than the actual values members act upon (Argyris and Schon, 1978). According to Schein (1984), "to really understand a culture and to ascertain more completely the group's values and overt behavior, it is imperative to delve into the underlying assumptions, which are typically unconscious but which actually determine how group members perceive, think, and feel."

In terms of DFM, when particular cultural values and underlying assumptions have evolved over the years to define the social worlds of the design and the manufacturing engineer, these socially constructed meanings are apt to endure even when management begins to espouse the philosophy of cooperation and customer orientation. Argyris and Schon (1978) provide many examples where "espoused theories" by managers contradict their "theories in use." Thus, while design management may seek to gain legitimacy (Meyer and Scott, 1983) from the corporate hierarchy by espousing a philosophy of "teamwork" and DFM, the actual rewards they deliver may follow more closely the old philosophy of a separation of the goals of product engineers and manufacturing process engineers and actually encourage individualistic, competitive behavior.

Closely linked to the concept of culture is that of career socialization. Culture endures by being passed on to new members. So newly hired engineers learn the norms, values, and underlying assumptions of the organization. As we will see, in one of the cases about 100 new engineers were hired after the DFM program was set in motion, providing a rare opportunity to socialize a new cohort of engineers to think and behave in a way more compatible with the new DFM strategy.

With this basic conceptual framework, we turn to the case examples.

CASE BACKGROUND

The U.S. automotive industry has often been criticized for its lack of innovation, particularly in comparison with its counterparts in Europe and Japan. In *The Productivity Dilemma*, Abernathy (1978) documented the slow rate of innovation and attributed it to a trade-off between productivity and change. Once large capital expenditures are made on inflexible automation, the argument goes, there is a conservative bias against product change. According to Abernathy, writing in the 1970s (p. 10):

> Major components of the automobile are manufactured in highly specialized and automated production plants; the scale of and capability of these facilities are now critically important in determining the types of changes that can be made in the product. Innovation has given way to standardization as a competitive tool; product diversity has given way to economies of scale; and external pressure on the industry has replaced entrepreneurial action as the major stimulant of technological change.

More recently, however, Abernathy, Clark, and Kantrow (1983) observed that the U.S. automotive industry in the 1980s entered a period of "dematurity" in which a reversal in automotive technology has generated rapid changes in vehicle design and manufacturing processes. This has been driven by competition from abroad, particularly from the Japanese, who are able to innovate more rapidly.

The recent movement in the U.S. auto industry toward DFM is driven by three major goals. First, DFM is seen as a way to reduce product development lead time. Clark and Fujimoto (1989a,b) have documented the U.S./Japan gap in lead time and found evidence that Japan's shorter lead time is due in part to the practice of "overlapping problem-solving." That is, through concurrent engineering, products and manufacturing processes are designed more-or-less simultaneously, in contrast to the serial approach used in the U.S., in which the product is designed and then handed off to manufacturing. Takeuchi and Nonaka (1986) liken the U.S. approach to a relay race and the Japanese approach to a rugby match. Second, DFM is seen as a way to increase product quality. There are indications that when the product is designed for manufacturability, defect rates drop. Third, all of this means reduced product cost.

AUTOCO very much fits the evolution described by Abernathy and associates. Prior to the late 1970s the company was highly bureaucratized, with distinct functional "chimneys" that operated according to their own standard operating procedures. Rewards were based on pleasing one's functional boss. Teamwork across functions was all but absent. Decision-making followed the chain of command and was highly autocratic. Reducing short-term costs was the main objective. The intense competition from Japan in the wake of the 1979 oil embargo had a profound effect on management practices. Top management instituted a set of corporate-wide quality and human resource programs, many of which encouraged greater employee participation and teamwork across functional boundaries. A new mission statement was developed by top management, with input from the union, to reflect this shift in emphasis from short-term costs to quality and the value of their human resources. The result of these efforts over the decade of the 1980s has been a major cultural change—a shift in values, operating philosophy, and beliefs. Compared to the traditional culture of the past, the emerging new culture seems considerably more conducive to DFM practices.

The two divisions we used for the DFM case studies in this paper were selected because of their substantial differences in size, level of innovation, and the contrast between a core production division and a component parts division. Auto Parts is a manufacturer of a variety of small component systems used throughout the vehicle. Engine is responsible for engine design and manufacture. At Engine, we focused on the product and process designers of the piston and rod assembly for the most recent engine program and visited the location where the new engine was recently launched. Much of the data focus on the design of the connecting rod, a machined part that connects the pistons to the crankshaft.

Auto Parts

In Japan the majority of automotive components are made by outside suppliers (70–75%), whereas U.S. automakers make more of their components in-house (40–50%). In the United States, internal component operations were historically treated as insiders—one of the limbs of the automotive company. Typically they were large-volume producers and were able to compete with outside suppliers because of their economies of scale (Liker and Kamath, 1991). Small-volume components were sourced outside. Thus, in-house component suppliers made large-volume parts that were very mature and innovation was quite limited (Abernathy and Utterback, 1978).

Much of the innovation occurred in the shops of smaller outside parts suppliers. The automotive companies often gave some business to these outsiders, who developed innovative designs and then licensed the design to be made in large volume in-house. This led to much antagonism between outside and inside part suppliers (Liker and Kamath, 1991). Outside part suppliers claimed that they had no choice but to license their designs (or even give away their designs) or lose any hope of doing future business with the automotive giants. Inside parts suppliers claimed that their relatively high negotiated labor costs gave them a disadvantage compared to outside suppliers.

This situation changed considerably in the 1980s. U.S. auto companies are attempting to emulate the closer, more cooperative relationships with outside suppliers that characterize the Japanese system (Liker and Kamath, 1991). They are making more long-term commitments to outside suppliers and involving them earlier in the design process. At the same time they are raising the targets for cost and quality for both their internal and external suppliers. Price is no longer the only criterion for selection of supplier—quality and technical design expertise are increasingly being given consideration. Moreover, internal suppliers are no longer given an automatic advantage in sourcing, increasing the competitive threat of losing business or even being closed down in favor of outside parts suppliers.

This general scenario fits Auto Parts quite well. Many of the components they make are based on modifications of very old designs dating back 20 or more years, and some of the more recent designs were licensed from outside competitors. Until recently Auto Parts did not exist but was part of another division. About three years before our visits Auto Parts was created, and some component designers were transferred to it, though their former parent division retained the authority for final release of designs. The goal is to make Auto Parts a wholly independent operation with its own design capacity. In the past two years AUTOCO has begun to change its strategy and view Auto Parts as a potentially profitable operation, though there is still some internal doubt about its long-term fate.

At the time of the interviews Auto Parts was still in the process of developing a full-scale product design organization. There was still ambiguity about what design responsibility they had as a division compared to what responsibility was retained in their former parent division, and there was a power struggle for new product design responsibility. The former parent division felt they had the greater expertise for new product design and that routine engineering changes could be best handled by design staff at the plants, whereas Auto Parts wanted to control their own future products. This is further complicated by the fact that Auto Parts was officially to be treated like an outside supplier that needed to be monitored by a separate internal engineering organization.

Those component design engineers who were transferred to the plant were assigned to the same department as manufacturing engineers. There was some cross-training of product designers and manufacturing engineers. In addition, approximately 100 new engineers were hired in the one-year prior to the interviews. Many of the new hires are in a corporate-wide program for fast-track engineers and began with three 8–month rotations between manufacturing and design. The long-term goal of the division is to create a combined role—individuals responsible for both product and process design. Thus, Auto Parts is viewed as a pioneer in their approach to DFM within the company.

Engine

Even more than Auto Parts, Engine was historically protected from outside competition. However, when the company decided to build their newest engine, they began by looking to a Japanese company as the source. The Japanese company already had a prototype engine, and they had proven cost and quality levels that surpassed anything AUTOCO had been able to achieve in the past. Nonetheless, Engine was given a chance to bid on the new engine program. They put together a business plan designed to match their Japanese competitors in cost and quality. Part of the plan was a new approach to human resource management that included the use of autonomous work teams in the manufacturing operation and the use of cross-functional teams to design the product and process concurrently. A central part of the concept was the use of a design approach that allowed multiple sizes and configurations of engines to be made on the same line. AUTOCO decided to take a chance on their own internal Engine Division, in part because of the strategic importance of the vehicle power-train.

The engine design was begun in 1985 and the first engine was produced in April of 1990. The organizational approach to DFM and the sociotechnical systems design approach of the manufacturing operation have been revolutionary for AUTOCO. Because of early signs of the success of the program—an on-time launch with few major problems, and the high quality and performance levels of the early engines—the approaches used here became the model for future engine programs at AUTOCO.

OLD WAYS AND OLD BARRIERS

We asked those interviewed in Engine and Auto Parts to describe the old design process, before there were any special DFM programs instituted. We were able to get both the design and manufacturing perspectives. Since the old design process was so similar across divisions we present here a single generic description.

The Old Design Process

A general description of the traditional design process by consultant Jim Harbour fits the old design process of AUTOCO well (1987, p. 23):

> Typically our design and product engineers develop the designs for new products, then a separate group of manufacturing engineers develop the facilities and the processes used to fabricate and assemble the same products. By the time these products enter the production plants, the design and product engineers are off working on the next major venture without another thought about the previous project . . . . Few, if any of these design engineers have ever spent a minute in a manufacturing plant.

A simplified version of the old component product and manufacturing design process at AUTOCO is presented in Figure 13.3. Manufacturing is not even informed about the product design features until the component design is being refined in the second iteration, based on prototype test results. As soon as manufacturing gets some indication of the design features, perhaps even obtains a blueprint, they begin to design their process and speak to equipment vendors who actually design the equipment. Design-manufacturing feasibility meetings are held, but these are mainly reactive—manufacturing lists their concerns about potential problems in manufacturing and assembly, and design engineering tries to accommodate them with minor modifications. The main issue is not how best to design the part so it can be manufactured at high quality and low cost, but whether manufacturing is even feasible. These meetings were described to us as adversarial. In response to manufacturing's claims that they could not build the part, product designers might argue that they have seen other manufacturers build it and accuse manufacturing of being in the dark ages technologically.

It also should be noted that, as Harbour observes, once the product and process

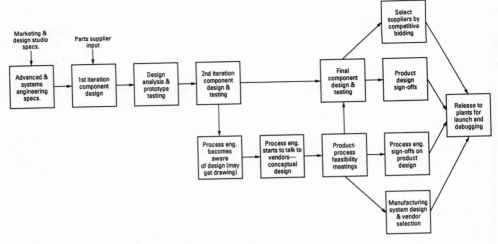

Figure 13.3. Traditional design process at Engine Division.

design is complete, and the manufacturing system is able to run a certain number of good parts, the product and process designers, both located in divisional offices, move on to other projects. Responsibility for making the part is then turned over to manufacturing engineers and other personnel at the plant. Division-level product and process designers will then provide support to the plant ''as needed'' (e.g., all product changes must be formally approved by product engineers who have sign-off authority).

Having described the overall process, we can now analyze the process in terms of the barriers created by the traditional system. We organize the barriers into the three general categories of Figure 13.1—formal organization, political, and cultural.

Formal Organization Barriers

With roots firmly in the Tayloristic principles of scientific management, the formal organization of design in AUTOCO was highly mechanistic. The process of design was divided into a large number of smaller parts that were delegated to a large number of specialists each with specific duties and authority. The total vehicle was divided into systems, subsystems, and individual components, each the responsibility of a specific group or individual with needs to span social, geographic, departmental, and company boundaries.

As Lawrence and Lorsch (1967) observed, when a process that is naturally integrated gets divided into tiny parts assigned to separate departments and functions, they must somehow be reintegrated. Not only do parts have to be designed so they are manufacturable, but different components have to fit with each other and into a specific envelope; e.g., the piston must fit the bore size in the cylinder block, and the engine must fit into the engine compartment. Galbraith (1977) suggests there are a limited number of approaches to integration, some of which build mechanisms for ongoing communication and coordination, thereby increasing the organization's information processing requirements (e.g., task forces) and others based on rules and standard operating procedures that reduce information-processing requirements.

The traditional design process, with its emphasis on efficiency, prefers to minimize information-processing requirements through the use of an extensive division of labor and specifications passed in linear, serial fashion. So, for example, typically at AUTOCO the car is designed from the outside in. The body styling is completed first, which provides the envelope into which components must fit. "Systems engineers" are responsible for major component subsystems, e.g., the exhaust system, and give broad specifications to engineers who design at the component level, e.g., the muffler. One can view the process as a project evaluation review technique (PERT) diagram, starting with broad general decisions by a centralized product planning group and branching out to a multitude of nodes at which engineers are making microscopic decisions about tiny parts of the car. At each successive stage, there is a hand-off of specifications from the previous stage, and communication across branches is minimal. So, for example, the muffler designer is not apt to communicate with the designer of the car underbody, even though these parts must fit together. The interface has been previously worked out through specifications. As we will see, this extensive division of labor across components continued with the new DFM program. For example, at Engine an entire cross-functional design team was working on the design of a connecting rod while separate design teams worked on piston design, cylinder block design, etc. Communication across teams was minimal.

A summary of the key roles involved in the design process is presented in Table 13.1. Each role sees only a small piece of the total design process. Performance of personnel in each "chimney" is measured in different ways, reinforcing the differentiated structure. In theory, this structure, with appropriate standard operating procedures, creates a set of checks and balances to prevent overemphasis on any one goal. For example, the advanced engineer's drive for innovation is checked by the component engineer's desire to avoid risky products that might increase warranty costs. The component engineer's desire to work closely with the design staff of potentially pricey outside parts suppliers to benefit from their engineering expertise is checked by the purchasing agent's desire to source at the lowest possible cost. The product engineer's emphasis on product performance is checked by the process designer's emphasis on manufacturing feasibility. And so on . . . .

In fact, as theories of bureaucracy would predict, this creates a highly inflexible system that stifles any innovation that might enhance product functioning and quality. None of the roles has an incentive to really meet the needs of their immediate customers—those who depend on their services are apt to be treated as adversaries. The lack of cooperation and coordination ultimately leads to an environment in which each function is trying to find its own "life preserver merely to stay afloat" (Schonberger, 1990).

A good example of the barriers to DFM created by this highly bureaucratic system is in the engineering change process. The engineering change process is one of the few opportunities for intensive communication between product and process engineers. Often a first assignment for a new product engineer is to respond to requests for changes from manufacturing. This experience is often the principal way in which a young engineer gets exposed to the manufacturing implications of design decisions. Unfortunately this opportunity for learning is limited by the bureaucracy,

Table 13.1. Key Functional Roles in Traditional Design Process

Functional role	Responsibility	Goals
Advanced engineer	Long-term product development	Product innovation
Systems engineer	Specs for automotive system	Component fit and overall product functioning to reduce warranty costs
Component (product) engineer	Detailed specs for component	Component functioning to reduce warranty costs
Product designer	Develop detailed design	Efficiently provide detailed design that meets specs
Drafter	Develop detailed drawings	Efficiently make drawings that meet drafting standards
Process design engineer	Design manufacturing process and work with tool & equipment vendors	Design a process that works and meet schedule
Plant manufacturing engineer	Implement, maintain and improve manufacturing processes in the plant	Reduce manufacturing costs and maintain production
Resident engineer	Design engineer assigned to plant to facilitate design changes	Authorize design changes to maintain production while maintaining product functioning
Purchasing	Purchase parts, tools, and equipment	Purchase to engineering specs at minimum cost
Tool & equipment vendors	Sales engineer is liaison to customer's process design engineer	Provide services required to get and keep business
Parts suppliers	Sales engineer is liaison to customer's component design engineer	Provide services required to get and keep business

which creates distance between the product and process engineers. An example of the approval process was provided by an employee of a different division than those studied—the division responsible for stamping and assembling the cars. This account from a former employee of AUTOCO is not of the distant past, but rather of experiences in 1989.[2]

The design change and associated drawings are sent from product engineering to the process engineering general office, one mile away. This bundle arrives on the process engineer's desk; he or she reviews the change, makes appropriate cost estimates for changing the tooling, and then forwards this bundle to their supervisor for approval. It is then sent to the next process engineer involved with the assembly of that part, and the cycle continues. Often more than a half-dozen process engineers must review the change. Although a product engineer may be confident that none of the others will be affected by the change, he or she cannot approve the change for them. The bundle of drawings must be forwarded to each individual and approved by their supervisor. The supervisor's approval is usually unnecessary as he or she has little knowledge of the intricacies of all projects and acts to slow the process. In the end, the approvals reach the product engineer long after the requested response date, which does not seem to concern the product engineer.

While the brick wall between product and process engineering is well known, it is less well known that process engineering is a brick wall away from the plant manufacturing engineers. A dynamic similar to the one that occurs between product and process engineers also characterizes the process and manufacturing engineering relationship. Process engineering is the responsibility of the division-level engineers located in central office locations. Engineers at the plants do not become involved until the launch, when responsibility is then transferred. Because of this division in work domains, process changes are often made after launch without the knowledge of those at the division level. Thus, documentation (division-level responsibility) often does not correspond to actual practice. The result according to the same former employee:

> Problems arise during quality audits when outside auditors cannot identify the welds according to documented weld numbers. The original tool numbers, stage numbers, and weld sequence are often changed by engineers stationed at the plant unbeknownst to the general office. Trips to the plants are then required to resolve the discrepancies. The plants sometimes make changes which have adverse effects on the part or process; they are of course unaware of this because of their lack of involvement up front in the process engineering phase.

In sum, the formal organization is designed for control and efficiency. Presumably, if everyone does their job according to standard operating procedures, communication will be streamlined, checks and balances will keep all parties in line, and the right specialized expertise will be brought to bear on design decisions. When we look below the surface of the formal organization, what goes on at the day-to-day level is better described by what has come to be called the "informal organization," which includes the organizational politics and cultural symbols that make up ongoing organizational life.

Political Barriers

Conflicting interdepartmental goals. The divergent goals of each of the major roles in the design process leads to game-playing and attempts to gain power. An excellent example of this game-playing was described by a process designer in Engine who was reflecting on the difference between the new DFM approach and the more traditional design process. He explained that a major difference between the design process for the DFM approach and the old approach was that now he lets the product engineer call his equipment vendor directly to work through DFM issues. In the old days "that would never happen or I would be on the phone to the product engineer immediately!" Why? we asked. He explained that they got their budget for process design from the product design side of the division (of which component engineering was a part) and they frequently inflated budget estimates of the cost of services provided by equipment vendors. In this way they had slack resources to deal with unexpected problems and process improvements that had not been approved. If the component engineers got too close to the vendors they might learn some of "manufacturing's secrets."

Product engineers have a great deal of formal authority, as they ultimately can authorize or veto product design changes requested by manufacturing. In addition to formal approval power, product engineers derive much of their power from controlling information flow. As Mintzberg (1983, p. 184) observes, an important basis of political power is "centrality," the power that derives from "standing at the crossroads of important flows of information." Product engineers have this power despite their youth and inexperience with the particular component for which they are responsible (as discussed below). They are dealing with process engineers, outside suppliers, and product "designers" who may have decades of experience with the component. Yet as a result of their position they control the flow of critical information. Only they have access to timely design decisions that impact the work of process engineers and that outside suppliers would love to know about to get a jump on the bidding process.

While product engineers have considerable power over product specifications, manufacturing personnel have their own sources of power. Since product engineers rarely visit the plants, manufacturing can get away with shipping parts that are not within tolerance. Component engineers then have to catch manufacturing by investigating warranty claims that might possibly be attributed to intentional deviations from specifications by manufacturing—something difficult to prove. When components are in production, there is tremendous pressure to get them to the assembly plants to keep the line running. Respondents described cases of manufacturing using two different gauges to test parts, one for their own use, which was to the tolerances they believed were acceptable to achieve a quality part, and a tighter gauge matching the blueprint to show product engineers that they only accept parts within the drawing's true tolerances.[3]

The product engineer responsible for the piston rod design at Engine described an excellent example of manufacturing's power from his last traditionally managed engine program. After the product and process designs were complete and the manufacturing process was being installed and tested, the product engineer had a material handling concern. It seems that the piston castings came in metal crates that were lifted by elevators and were being dumped from the crate into a hopper. The product engineer was concerned about damage, in this case bent piston castings, that would affect engine functioning. He wrote a letter, with supporting data, requesting the process be changed, but no action was taken. When it came time for him to sign off on the manufacturing process, he signed with a contingency that they change to loading by hand or some other method. This did not stop manufacturing. As of this writing they continue to load parts by dumping castings into a hopper. Product designers, with bittersweet humor, refer to this process as the "demon drop" after a roller coaster at a Midwest amusement park.

Conflicting customer-supplier goals. The process of interacting with vendors and suppliers has its own set of interorganizational politics. As described above, parts suppliers and equipment vendors are an integral link in DFM. Parts suppliers are often the main source of real product engineering expertise and the automotive company's component engineers often act primarily as coordinators working with suppliers. The company's component engineers may in fact lack the product knowledge to do any of their own original design. On the manufacturing equipment side, automotive companies generally don't design manufacturing equipment per se. While they usually design the overall process, they typically rely on equipment and tooling vendors to design the actual manufacturing equipment. The main equipment vendor usually acts as a system integrator, much as a general contractor serves in construction. Thus, like their component engineering counterparts, process engineers act as coordinators working with outside vendors.

If we step back from this process we can see that much of product design and process design is taking place at the shops of outside vendors and parts suppliers. Thus, effective concurrent engineering may require early involvement of the product engineers of parts suppliers and the manufacturing engineers of equipment vendors, as is more commonly practiced in Japan (Liker and Kamath, 1991). What often prevents this is the traditional purchasing process discussed above.

To recap, since purchasing may refuse to source based on any criteria but cost, it is quite possible that outside vendors and suppliers, who invest many engineering hours working with AUTOCO engineers to develop designs, find that they are not given any of the business from purchasing.[4] In fact, the business to make the product they designed might go to a competitor or even a division of AUTOCO that lacks any sophisticated design expertise (Liker and Kamath, 1991). One way to combat this appropriation of design ideas is to hold back key information until after the contract is let. This might in fact be key manufacturing information about how to optimize the manufacturing process to build the particular design. Liker and Kamath (1991) interviewed some parts suppliers who explained that they refused to let engineers from the major auto companies in their plants because they did not want to give away their competitive manufacturing secrets. The implications for DFM are obviously quite serious.

Conflicting career/DFM goals. As Morgan (1986) observes, individuals have interests in furthering their own careers that often come into conflict with interests of the broader organization. When component engineers are hired by AUTOCO, they quickly learn that the path to climbing the corporate hierarchy and achieving the highest possible salary is to get onto a management track as soon as possible. In fact, a select group is placed on a fast track to management when its members are first hired. Their early experiences in product design, and in some cases short rotations through manufacturing, are viewed as formative experiences—a training ground for management.

The main feedback that product engineers get about the success of their designs comes from warranty claims and requests for engineering changes from a variety of sources, including manufacturing. This feedback is most effective if the product engineer continues to work on the component he or she designed for at least one year after the design is released and then designs another component using the knowledge gained. This is not the case for many fast-track young engineers. A major impediment to the fast career progression of a component engineer is to become overly associated with a particular component. If they become known, for example, as a crankshaft designer and are repeatedly assigned to crankshaft design programs, their career progress might slow to a crawl. A rule of thumb is that three years is the maximum the young product engineer should spend in any one job. This happens to be just about the time it takes to complete a component design. So by the time their component is in production, and before there has been enough time to get warranty data, the young engineer has moved on.

The conflict between developing high levels of technical expertise in organizations and selecting highly qualified technical managers has been long recognized in literature on the careers of technical professionals (Badawy, 1982). What has not been discussed are the implications for DFM. This system rewards generalist knowledge at the expense of in-depth specialist knowledge about a particular component. DFM depends on this specialist knowledge in two ways. First, much of the specific knowledge about the best way to design a part to optimize on both product performance and manufacturability is probably learned through experience over several design cycles. If component engineers are transferred after their first major experience designing a part the organization loses the capacity to accumulate a base of component-specific experience. Second, part of what comes with experience is the development of a network of social contacts with process engineers responsible for the component. The fleeting relationships that develop between design and manufacturing in a system of constant movement of engineers are not apt to lead to the trust and sharing of information and knowledge that characterizes the kinds of DFM processes described in this book.

The career system we have described means that many product engineers lack the in-depth knowledge and experience actually to design the components (Liker and Hancock, 1986). So how do parts ever get designed? In Table 13.1 we note the role of product designers who do the detailed design work. Product designers are former drafters, without a college degree, who through experience were elevated to a position of greater responsibility. They develop the detailed designs based on rough sketches and specifications provided by product engineers. An experienced product designer might be given a great deal of latitude by the product engineer to make major design decisions. Since designers do not rotate through different components as rapidly as product engineers, designers become the stable source of in-depth component knowledge.[5] These individuals often have a decade or more of experience with a particular component. While they are more experienced with the component and its manufacture and have the capability to design a complete part, they do not have the authority to release designs and they depend on the component engineer for the specifications. This is not because the component engineer is necessarily technically better able to understand or develop the specifications, but because the product designer sits at a drawing board (or more recently a CAD terminal), while the component engineer is in the center of key communication flows and has formal authority over the design specifications. Senior product designers often complain bitterly about the power of young component engineers who they claim have only the advantage of a freshly printed college diploma.

Thus, the division of labor extends to a distinction between the component engineers, who act as managers of the flow of design information and have authority over design decisions, and the designers, who actually create the design. Although component engineers often lack knowledge of manufacturing implications of design decisions and have not had time to establish close relationships with manufacturing, it is they who have authority over design decisions and they who typically are assigned to cross-functional teams to bring together product and process expertise. This raises the interesting question of whether DFM can proceed effectively without the direct participation of designers or without revamping the career system so that component engineers have incentives to develop in-depth component expertise.

Cultural Barriers

Incompatible product-process engineering subcultures. The different subcultures of design and manufacturing at AUTOCO symbolically reinforce the conflict that results from concrete goal differences between these groups. As Morgan (1986, p. 127) notes:

> In organizations there are often many different and competing value systems that create a mosaic of organizational realities rather than a uniform corporate culture. For example, different professional groups may each have a different view of the world and of the nature of their organization's business . . . . The frame of reference guiding development engineers may be different from the perspective of members of the production department.

The cultural differences between product engineering and process engineering at AUTOCO were dramatic and could be seen at all three of Schein's (1984) levels of culture. At the level of visible artifacts, there was uniform agreement among respondents that product engineering was the higher-status "glamour job." Product engineers had a college diploma; worked in clean offices away from the noisy, dirty plant environment; and interacted with other professionals. There is virtually unlimited potential for rising in the management hierarchy. By contrast, manufacturing engineers often come from the ranks of the skilled trades and generally cannot rise above middle management within a plant. At best they can get a promotion to division-level process engineering, which takes them into an office environment that is almost as good as that of design engineers. Product engineers live in a world of paper, abstract concepts, and meetings, while manufacturing engineers live in a world of machines, tooling, and action. While product engineers speak the language of engineering analysis and test results, manufacturing engineers speak the language of machining processes and manufacturing efficiency.

The cultural barriers between process engineering, located in central offices, and plant personnel can be as great as those between product engineering and process engineering. One process engineer in Engine had worked his way up to division process engineering from the ranks of an engine plant. He described his early experience working with plant personnel after his promotion to division in this way: "Plant people thought staff people didn't know anything. Here I was working with the same people I had worked with for years in the plant, but now I was a staff guy. I had to say at one meeting, 'When I went to staff did they give me a lobotomy and take away what I know?' "

The value systems of product design staff and manufacturing staff reflect their different goal orientations discussed earlier. Product engineers value a highly functional product that is elegantly engineered. There was also some indication that they value a highly technical process used to develop the design. For example, they spoke proudly about using sophisticated engineering analysis methods rather than relying exclusively on the older method of prototype testing. By contrast, manufacturing staff value a manufacturable product that works. They are suspicious of very complex designs or design processes and would prefer sticking to the tried and true. To manufacturing staff, if they cannot build it and see and feel it, it is not real. Thus, we heard reports of manufacturing impatience with participation in conceptual design discussions before a prototype was built. The manufacturing engineer takes pride in being close to the means of production—the bread and butter of the company—and is somewhat disdainful of staff who seem a step removed from the "real world" of actual production.

An important underlying assumption of both product and process designers is that product design and manufacturing are two separable and discontinuous processes. Entire social worlds and formal organizational structures have been built over the years to reinforce this assumption. One can work in product design or manufacturing but not both. Working in one of these activities means having a different base of experience and formal credentials, a different office environment, dressing differently, speaking different languages, and having different formal responsibility. Breaking down the brick wall between design and manufacturing may mean breaking down this taken-for-granted assumption that there is actually something real in the division between the two sets of activities.

Low status of manufacturing. Traditionally, design has been essentially a class system in which product engineers are the aristocrats and manufacturing engineers the supervisors of peasants at work. The engineering degree is a credential that separates the elite product engineers from those manufacturing "engineers" who still get their hands dirty, having made their way up from the working class. This is the case even though the elite design engineers rarely use the high-level analytic tools they were taught in college. The status advantage of product engineers over manufacturing engineers at AUTOCO was reinforced by the fact that their job was rated a grade level higher.

Individuals in design and manufacturing come to the job with different skills and expectations—product engineers are the fast-track college kids, and manufacturing engineers are the nuts-and-bolts guys that got a break from blue-collar work. There is little cross-over in career paths and thus little common experience or expertise. For a product engineer to spend time beyond a short rotation early in his or her career in manufacturing could mean falling off the product engineering career path. Manufacturing staff develop a set of values favoring working on the front line that helps insulate them from the frustration of stunted career possibilities. Yet they cannot totally ignore the fact that they are not likely to achieve the career heights and associated status perks available to the product engineer. Thus, status differences serve to add yet another barrier to trust and open communication between the product and manufacturing process sides of the business.

Individualistic values. The competing interests across functions, conflicts between career aspirations and organizational needs, and the status hierarchy of design and manufacturing all reflect a highly competitive climate within the corporation. This culture, which emphasizes individual competitiveness over teamwork, is also a reflection of American culture more generally. One of the more poignant commentaries on the costs of the individualistic value system in contemporary America is provided by Bellah and colleagues (1985) in their book *Habits of the Heart*. Analyzing society as a whole, they write of the fragmentation and lack of coherence in our contemporary culture. The lack of a sense of community and lack of dedication to goals larger than the individual's personal interests can be seen inside and outside of work. The American ethic of competitive individualism reinforces the individual's desire to be a winner and to be part of a winning team. As Morgan (1986, p. 119) observes: "From an American perspective, industrial and economic performance is often understood as a kind of game. And the general orientation in many organizations is to play the game for all it's worth: set objectives, clarify accountability, and 'kick ass' or reward success lavishly and conspicuously."

What we have described in this highly differentiated organizational structure are a set of competing loyalties and thus competing bases of identification. Should engineers try to win for the company as a whole, their functional chimney, their particular component team, or their own career advancement? Many of the DFM programs presume a cooperative spirit across organizational boundaries and an identification with the company as a whole. In theory it may be possible to align interests such that one can "win" at individual, departmental, and company levels simultaneously. However, short of realizing this ideal, a cooperative, team-oriented value system may be a necessary part of the culture supporting DFM.

THE NEW WAY: BREAKING DOWN BARRIERS

There have been many DFM efforts throughout AUTOCO. Different parts of AUTOCO have the autonomy to choose their own approaches, so there is a good deal of variation in how DFM is managed across the company. In this section we contrast the approaches to DFM taken by Engine and Auto Parts. Table 13.2 summarizes the organizational barriers discussed above and the actions taken at Engine and Auto Parts that seem to most directly impact these barriers. We discuss these actions below in greater detail.

Changes in Organizational Structure

The new design process used by Engine for the development of their latest engine is shown in Figure 13.4. Under the old system product and process design were in two separate chimneys, reporting up through separate hierarchies that first came together at the level of vice president. The first major change was to create a new product and manufacturing engineering organization with a director two levels closer to the actual engineers responsible for design. The second major change was the movement of process engineers from their building one mile down the road to the product design building. The third key change was the use of cross-functional teams, as shown in the concurrent engineering design process in Figure 13.4, which required new forms of contract relationships with parts suppliers and equipment vendors.

Table 13.2. Organizational Context Barriers to DFM and Actions Taken at Engine and Auto Parts

Organizational barriers	Engine actions	Auto Parts actions
Formal organization		
Mechanistic organization (high differentiation, rigid boundaries, high control orientation)	Merge product-process reporting lines	Merge product-process reporting lines
	Co-locate product-process engineers	Co-locate product-process engineers
	Cross-functional teams	Combine product-process engineering dept.
	Process designers implement own designs as part of autonomous work groups	Cross-train product-process engineers
	Resident product engineer in plant	
Political		
Conflicting interdepartment goals	Merge product-process reporting lines	Merge product-process reporting lines
Conflicting customer-supplier goals	Early component supplier sourcing; early equipment engineering contract	Limited action
Conflicting career/DFM goals	No action	Future plans for combined product-process engineering roles
Cultural		
Incompatible product/process engineering subcultures	Co-locate product-process engineers	Co-locate product-process engineers
	Cross-functional teams	Combine product-process engineering dept.
	Geometric tolerancing training & use	Cross-train product-process engineers
Low status of manufacturing	Raise manufacturing pay grade	Raise manufacturing pay grade
		New recruit socialization
Individualistic values	Emphasize teamwork as corporate value	Emphasize teamwork as corporate value
	Teamwork training	Teamwork training

The first, division-level, cross-functional teams at the level of component design and process design were established in the conceptual design stage of the new engine program and stayed together until the component design was complete. The connecting-rod team was led by a component engineer. Members included a manufacturing process engineer, a manufacturing assembly engineer, statistical quality assurance, supplier representatives, designers/drafters (who did the actual design work), material control (who specified the part numbers for prototype engines), and purchasing (who

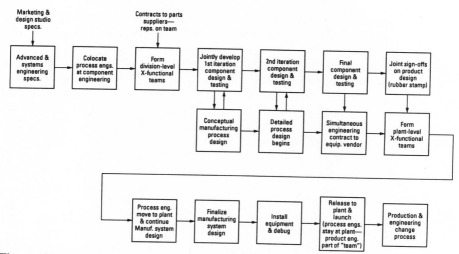

Figure 13.4. Concurrent design process at Engine Division.

did the actual buying). There were periodic scheduled meetings where practically all members attended, but most often the component engineer invited only members with relevant knowledge for particular decisions. The suppliers were selected early by product engineering, process engineering, and purchasing and awarded a long-term contract. Equipment vendors were not yet selected at this point.

The second, plant-level, cross-functional teams were established to develop the detailed manufacturing system design and supervise the production launch for each major component system and assembly. The main players in these teams were product engineering, manufacturing, machine tool vendors, the production team leader, the hourly production coordinator, and support (e.g., a financial analyst). Machine tool vendors were not guaranteed the equipment supply business, but they were given a "simultaneous engineering" contract, a recent innovation for AUTOCO. This paid for their up-front engineering time.

Continuity between the component design process and the manufacturing design process was provided by the transfer of the process engineer from the division level to the plant level. Typically at AUTOCO when process engineers completed the design of the manufacturing process for a plant they moved on to another design project, as Harbour (1987) observed. In this case, process engineers continued as employees of the plant through the launch and into production when they became part of autonomous work groups in a production unit. To involve actual production people, the production team leader and a new manager, an hourly "production coordinator," were made full members of the launch team about two years before the plant opened to participate in equipment design, selection, and plant launch. A product engineer, given the title of "resident engineer," was assigned to the plant through the launch and continued into production to "bring the cross-functional team members back together when they break down."

Auto Parts had not recently designed a major new product and process from scratch, and therefore they could not describe an entire product-process design under the new DFM system. However, they had made major DFM changes that they expected to have important effects on future products and that were already having effects on product and process modifications. As did Engine, they changed reporting lines so product and process design reported to a chief engineer, several levels below where product and process design previously linked up. The core feature of their DFM program was permanent co-location of product and process engineers from a central division office to the main manufacturing plant of the division. Product design and manufacturing engineering departments, previously separate, were merged into single departments with a supervisor responsible for product-process engineering. The ultimate goal in this division was to combine the product and process design roles to become the responsibility of single individuals, a radical step for AUTOCO that placed Auto Parts at the forefront of company attempts to design for manufacturability.

Changes in Organizational Politics

The DFM program at Engine was more successful in breaking down many of the old political barriers because it was a new product design to be built in a new plant. Auto Parts also made progress, but more vestiges of past political barriers persisted.

Conflicting interdepartmental goals. The key change that respondents felt reduced interdepartmental conflict between design and manufacturing was the merging of product and process reporting lines. Product and process engineers found themselves in meeting rooms jointly giving presentations to their common boss. He did not want to know only what product designers were doing for functionality or what process engineers were doing for manufacturing efficiency, but rather how were they working together to optimize product and process goals.

In both Engine and Auto Parts we repeatedly heard that the DFM programs had blurred the boundaries between product and process engineering. Product engineers better understood manufacturing concerns, e.g., the manufacturing implications of setting very tight tolerances. Process engineers better understood the functional requirements of the product design. However, we also learned that product and process engineering still retained their own goals as primary. As explained by a connecting-rod designer at Engine, "You can never get to the point where product and process have common goals. Product function is the key to me, more so than manufacturability. At the plant, making parts at low cost with low rejects is more important to them than light weight, fuel economy, etc."

There was some lingering interdepartmental conflict holding Auto Parts back from achieving their ambitious DFM goals. The conflict was not so much between product and process engineering, but had to do with Auto Parts' earlier roots as part of the larger parent division.

First, final sign-off authority on products remained in the former parent division, where a large number of product engineers who designed the components made by Auto Parts remained. Although Auto Parts was an internal parts supplier, it was presumed that they should be treated like an external supplier and needed to be controlled by a more central division of AUTOCO. While the plan was to shrink the product group in the former parent division that was responsible for Auto Parts designs, they were actually expanding. There was continuing tension between the two divisions over design responsibility. Auto Parts engineers perceived this as empire building by the former parent division. Second, much of the drafting and laboratory test capability remained in the former parent division's location, so Auto Parts could not act as a full-service design organization at their manufacturing plant location. Thus, we see in this case that the conflict shifted from tension between product and process engineering to tension between the old product engineering group and the new merged product-process engineering departments.

Conflicting customer/supplier goals. We rarely have seen mention of the role of outside suppliers and vendors in literature on DFM. Yet, as discussed earlier, they are often the real source of design expertise for component parts, tooling, and equipment.

According to the connecting-rod designer at Engine, a large part of the success of their cross-functional teams was due to the role of the parts suppliers who would manufacture component parts (e.g., the casting supplier) early in the design process. This required a major change in standard bureaucratic procedures, a shift to early sourcing, and a long-term contract. He explained,

> The new purchasing philosophy is to try to make the suppliers part of the team. You take care of them and they will take care of you. We know and they know that there are future programs. The big suppliers are in the business for the long-term and want to prove themselves. That is happening in the next engine program. The guys we are happy with on this engine program don't have to compete for the business.

This radically new purchasing approach in the U.S. has been standard procedure in Japan for years where large auto companies have a family of first-tier dedicated parts suppliers to whom they make a long-term commitment (Liker and Kamath, 1991). We seem to be just discovering the advantages of this approach in terms of DFM and quality. In fact, the weakest point of the connecting rod cross-functional team, according to the component engineer, was the lack of early manufacturing equipment vendor involvement in the product design stage. As shown in Figure 13.4, while the parts suppliers were put on the cross-functional teams early in the design process, the equipment vendor was not brought in until the detailed manufacturing system design stage. A lingering result of the historic separation of product and process engineering is that product engineering has purchasing departments to select parts suppliers, while process engineering has separate purchasing departments to select equipment vendors. Product engineering worked through their purchasing department to get an early contract to parts suppliers in this case, but process engineering decided not to pay the equipment vendor to become part of the original connecting-rod cross-functional team despite the fact that a single vendor designed and manufactured almost all of the equipment to make the connecting rod. When we asked the product engineer responsible for the cross-functional team whether he had any regrets about the process, he said that his major regret was the lack of early involvement of the equipment vendor.

Auto Parts' ambitious DFM program did not include involvement of parts suppliers and equipment vendors. They were only beginning to make changes in sourcing agreements with their equipment vendors, and they had adversarial relations with other outside suppliers with whom they competed for business. Yet, these outside competitors often had designed the parts that they were licensing and producing. As one Auto Parts product engineer complained, "They are making (product) design improvements and not telling us about them." It is interesting to note that Auto Parts were themselves a parts supplier that happened to be owned by AUTOCO. In their conflicted relationship with their former parent division they faced many of the problems of lack of design autonomy historically faced by outside suppliers to AUTOCO.

Conflicting career/DFM goals. Despite all of the DFM inspired changes in organizational structure, the career ladder at AUTOCO still favored engineers who had more breadth than depth and those who shifted into management. Young product designers explained that they had learned through the grapevine that you should stay at least 18 months in a particular area but that you should never stay longer than three years. "If someone gets good at one thing, AUTOCO is less likely to reward them with a promotion. If your name is associated with a particular product, then you are stuck." In this case Auto Parts was still very much influenced by the career system at AUTOCO as a whole. In fact, for many young product designers Auto Parts was viewed merely as a first step in their career path.

The connecting rod designer at Engine explained that he already had stayed too long in this area. As he was approaching the release of the connecting rod design for the new engine program, he turned down an offer for promotion to a supervisory position, which had set him back a salary-grade level, but he explained, "I really love design. I am not ready to quit designing yet." If he had accepted the offer, the company would have lost the continuity of his experience and the personal relationships he had established with process designers at the plant. The experience and personal relationships were important to Engine's DFM program for two reasons. First, the new engine plant was still in the early production stages, and there was a continuing need for product refinements that were facilitated by his intimate knowledge of the product and relationships with manufacturing people in the plant. Second, by staying a component engineer he was able to transfer his technical knowledge of DFM and his newly acquired organizational skills to the next new engine program.

The movement of process designers to the manufacturing plant they had helped design was a major change in their anticipated career path. In fact, a new career path was created by this move, the consequences of which were highly uncertain. Process engineers that we interviewed in the plant expressed some concern about their future. Since they were removed from the mainstream of process engineering in the central division offices they feared losing their centrality in the process engineering network. Managers might not think of them when challenging new projects or promotion opportunities arose. Some process engineers had managed to avoid being transferred to the plant. Others viewed their stay in the plant as short term and expected to transfer out within one or two years.

Changes in Organizational Culture

Incompatible product-process engineering subcultures. There was general agreement that bringing together product and process engineers to work as a team had given them each some appreciation for the other's world. At Engine, the physical movement of process engineers to the product design office was seen as a pivotal change, despite the fact that the process engineers had previously been located only one mile down the road. According to the connecting rod designer.

> They were not outsiders in our building anymore. They walked through drafting just like I do. I didn't even have to pick up a phone. I just ran into the manufacturing people. A lot of the informal one-on-one meetings were the most important to designing the product for manufacturability.

By transferring process designers to the manufacturing plant, cultural barriers were broken down between process engineers, who normally work in a separate office building, and plant personnel. "Getting people on the same team and forcing them to work together breaks down a lot of barriers. We knew we had to report as a team to the operating committee. We didn't just go back and report to separate bosses. We were baring our soul in front of a group of people."

Another interesting way of breaking down language barriers at Engine was through common training in geometric tolerancing. Geometric tolerancing provides standard ways of representing tolerances on blueprints that include information of the location of jigs and fixtures to be used in manufacturing. By training cross-functional team members in this tolerancing method, a common language concerning design was provided, which led to the surfacing of very explicit decisions about how to design parts so they could be effectively located on manufacturing fixtures.

A striking trend in our interviews was the general willingness sharply to criticize past design practices of the company. The old chimney structure, bureaucratic procedures, close-mindedness of product and process engineers, conflict, and lack of concern about quality were continually criticized. Part of the shared interpretation of the new social reality of AUTOCO was the retrospective critique of past practices. This provides an indication that a real cultural shift is taking place. What was valued as "good" under the old management system is now viewed as "bad," and an emerging new value system views cross-functional collaboration, teamwork, and quality for its own sake as "good."

The only exceptions we encountered to this generalization of changes in outlook were two older engineers who glorified the past and critiqued the present DFM emphasis. One of these engineers resented the implication that only recent hires in product engineering were trained to be sensitive to manufacturing issues.

> I've been here for 27 years. I never just threw the design over the wall. I designed manufacturing processes for ten years before becoming a product designer. These new engineers with high grade-point averages can't even read a blueprint. These young kids don't have any curiosity. They pick engineering because they want to become managers. We have students who can't take a bike apart. We hire 4.0 (gradepoint average) students and they have no common sense. They're into the books and refuse to get their hands dirty. They are more career oriented than product and company oriented.

It might be said these older engineers simply never bought into the "new culture," since they were part of the "old culture" that they helped create. However, evidence supporting the validity of some of their concerns about the "career orientation'' of the new engineers came from the new engineers themselves, as described above. While the recent hires were swept up by the new team concepts and language of DFM, they were not willing to commit more than a few years to learning about a single product line.

Low status of manufacturing. In both divisions manufacturing engineering pay grades were raised so that they were in line with those of product engineering. Employees of AUTOCO were very conscious of grade levels. They explained that under the old system it was difficult for a manufacturing process engineer to critique the design of a product engineer who was several grade levels higher. With salary grades leveled, product and process engineers could meet on equal footing. This would also undoubtedly smooth the way for Auto Parts' plans to create a combined product-process design role.

Auto Parts had a tremendous opportunity to socialize their new recruits into a new way of thinking about manufacturing. Auto Parts had hired about 100 engineers fresh out of school recently and thus had a great opportunity to influence their perceptions of manufacturing. In fact, 25% to 30% of their engineers had been hired in the past two years. Managers were quite proud that they were able to get many of their first choices and "capture 25% of those interviewed," despite the fact that these engineers were coming to work in a manufacturing environment. The employee relations manager explained that "it is harder to do a lot of little things here that make engineers feel like part of the engineering community. The central offices are analogous to a college campus. There is more advanced engineering work there and a longer-term product and process focus."

Two relatively young product engineers who had started out at the central offices explained that when they were first informed they would be moving to a manufacturing site they were hesitant: "I was not crazy about the idea at first. I didn't know what to expect. I initially had a strong bias against manufacturing. But after working in the central offices and realizing that anything I did, someone at the plant had a feasibility issue with, I realized the benefits." These product engineers explained that their attitudes toward manufacturing had changed dramatically since being at the plant. They had a growing appreciation for the challenges of manufacturing and for the concerns of manufacturing personnel. On the other hand, they also admitted that they saw their future in product engineering and in the long term in management. Neither wanted to get "stuck" in the plant long-term. One engineer explained: "Product designers have a greater opportunity to be creative. Product design is the focal point."

Individualistic values. The cross-functional team approaches in both divisions were very much supported by changes in the overall corporate culture, which was increasingly placing a high value on teamwork. These new values, whether "espoused theories" or "theories in use" (Argyris and Schon, 1978), were in sharp contrast to the old competitive individualistic value of AUTOCO's past. The value placed on teamwork was reinforced by numerous training programs that taught interpersonal skills and team problem-solving, as well as by managers who were rated according to how they worked as a team member. In fact, at the new engine plant they coined a phrase "teamability," which referred to an individual's ability to work as a cooperative team member. A low rating on teamability was purported to limit opportunities for promotions and pay increases.

In the old culture a meeting was a necessary evil—an activity that burned up valuable time and took engineers away from their "real work." With few exceptions those interviewed spoke glowingly about the cross-functional team meetings held for DFM purposes. A connecting-rod component engineer described "the key event" in this multi-year design project as a multi-day cross-functional meeting where the first component blueprint, ink barely dry, and process illustrations were placed on the wall:

> The drafter had comments about the way the feature dimensions had possible multiple interpretations, and we clarified the dimensioning. The outside tooling designer had key input about alterations to several features that would help him control tolerances better. In the past, by the time he would have made recommendations the parts would have been made and tested and we would be through two iterations. Unless there was a real problem that required change we wouldn't have made the change based on his recommendation. The biggest change was the use of geometric tolerancing suggested by the manufacturing guys. They said instead of designing a flat part of the rod sitting on a surface, we should show three points because this is more stable and keeps the part from rocking when you machine it.

Accomplishments

We have already discussed many of the benefits that our respondents claimed resulted from DFM. In the new engine program, the engine was launched on time, within budget (for the first time in 20 years), with few major crises; and early production engines are testing out as the best engines the company has ever produced. In the first three months of production, approximately one engine in a thousand developed problems on the road. Overall, quality of the engine is the best of any launch in the last 20 years at AUTOCO as measured by customer complaints, assembly plant concerns, and warranty experiences. We also asked about lead time, and the answer was that product and process lead time were unchanged. The reason—much of the lead time is spent waiting for the purchase of prototypes, the completion of tests, and the purchase of equipment.

The most concrete accomplishments that the Engine engineers could point to were specific product design decisions that took manufacturability into account. We have mentioned a few. Another example had to do with the design of the engine for flexible production. Since the product and process were being designed flexibly to make a variety of engine sizes, it was known that at least three different-sized connecting rods were to be built. Originally, the component designer sketched in three different-sized bolts—the smaller bolts for the smaller engines. When the outside supplier in a cross-functional team meeting explained the price advantage of using a common bolt and the process engineer explained the manufacturing efficiencies, the component designer did some analysis to determine the cost in weight and the stress properties of using one-size bolt. The results showed that product functioning would not be sacrificed and they went with one size. They then developed a list of all parts and dimensions and jointly rank-ordered the desirability of keeping these common across the three connecting rods. These became the criteria for many design decisions.

Since Auto Parts had yet to design a complete new product since the DFM program was instituted, and the total relocation of personnel and facilities was incomplete, most of the benefits they might achieve were yet to be realized. However, their experiences demonstrated that the advantages of DFM are not confined to the design of a new product. The continuing need for product and process refinements, in the launch stage and well into production, depends on continuing cooperation between product engineers and process engineers. Now, instead of manufacturing engineers throwing requests for engineering changes over the wall to product engineers, who throw an answer back, Mary calls her friend Joe with whom she has worked for years and they discuss the need for change. For example, product designers on one product line explained that they were licensing all of their designs from a foreign company and using their drawings. Their job was to convert the drawings to AU-TOCO format and then work with manufacturing as a "big launch team" to create the new manufacturing process to make the parts. The advantage of this new cooperative environment, according to these designers, was that the launch went smoother because they worked on the same team. The design people understood manufacturing's needs and could facilitate needed product changes, and manufacturing understood the implications of their decisions for product functioning.

In another example, when an automated fuel pump line was brought in, Auto Parts began with an existing fuel pump design. A cross-functional team worked together and vendors were brought in at various stages, which reduced finger-pointing and facilitated faster, higher-quality decisions. Although the product design didn't change, product knowledge was seen as crucial for decisions on how to handle the part so it would not be damaged and how to design a manufacturing system capable of adapting to future, evolutionary product changes. A major outcome was a shift from state-of-the-art manufacturing systems to less complex equipment that could make the part more reliably.

An excellent synopsis of the advantages of DFM to date at Auto Parts was provided by a chief engineer:

> So what are the results from the effort to reorganize for simultaneous engineering? The full effectiveness of the reorganization and relocation cannot be realized until additional drafting and laboratory test capability are provided at our plants. Already, engineers report that important decisions regarding design or process changes to their components are made more quickly, informally, and with better (firsthand) data."

Remaining Barriers

We have mentioned a number of continuing systems barriers to DFM. These included the lack of early equipment vendor involvement in the engine program; the reward system, which encourages engineers to rotate frequently across product lines and treats product design activities as a training ground for management; and the remaining changes required at Auto Parts to become self-sufficient in design capabilities. The example from Engine focused on a relatively affluent division that had been given a $2.5 billion budget to develop a major new engine program, the first phase of which cost $1 billion for engine development and plant renovation and launch. They started with a large budget and a clean sheet of paper, which gave them the slack resources to get outside of the existing bureaucracy and make an impressive set of changes in fundamental organization systems. On the other hand, we should note that not all teams functioned as effectively as the connecting-rod team. Moreover, according to the product engineer heading up the connecting-rod team, the lack of equipment vendor involvement early in the product design process was a major weakness. In this case, the bureaucratic separation of parts supplier purchasing and equipment vendor purchasing did not give the product engineer the authority to deal directly with equipment vendors.

Auto Parts had ambitious goals and made impressive strides in changing their organization. However, their experience is illustrative of the kind of self-reinforcing cycles that characterize bureaucracies (Merton, 1968; Selznick, 1949). Despite the commitment of their management and the substantial changes made, they were still quite entrenched in the existing bureaucracy of purchasing, the conflict over design responsibility between their new division and the former parent division, and the conflict between long-term product development and short-term support for current manufacturing. Resources continued to be allocated according to short-term priorities of production. A major complaint by engineers about their DFM approaches was that they lacked the resources—both personnel and the money—to mount a new product development program seriously. According to the oldest product engineer with whom we spoke, who was assigned to maintenance of existing product designs: "They formed a skunkworks in the plant, but didn't hire any extra people. They robbed us of product guys. They couldn't afford to put any manufacturing guys on the program." Thus simultaneous engineering became traditional product engineering. Another product engineer assigned to a "simultaneous engineering" program for a different product line told a similar story:

> The project lacked some engineering resources initially because the manufacturing engineers were busy with a production launch. By the time manufacturing was able to get involved, it was too late in the program for a simultaneous engineering agreement with the equipment vendors. We were constrained by program timing to release the design so that we could prepare our project proposal and submit for approval. As a result, design changes driven by manufacturing and the equipment vendors will now have to be implemented after project approval and may delay delivery of assembly equipment.

One engineering supervisor of a combined design-manufacturing department in Auto Parts contrasted AUTOCO's approach to product development to what she learned on visits to Japanese companies. Japanese companies put their money up front, bring the product to market, and believe that if they do quality work they will make a profit. In this company the financial aspects come very early in the development process. It is a cultural problem. We are not used to spending development money up front without clear-cut cost justifications.

THE ORGANIZATIONAL CONTEXT OF DFM: A SYNOPSIS

The analysis above used illustrations from two divisions of the same company. However, from our broader base of experience we believe the barriers discovered in these case studies are not peculiar to this one company. If anything, this is one of the more forward-thinking companies we have observed, and it has been unusually successful in making changes in its basic organizational structure and process to support DFM. Based on these experiences we would predict that the organizational context will have a significant impact on the success of DFM programs. We began this chapter discussing systems theory, which emphasizes the notion of "fit." That is, there is no one best way, but there must be an appropriate fit between different aspects of the organization and its environment. So, for example, in a highly mechanistic organization it seems unlikely that cross-functional teams will flourish and solve the problem of integrating different functional perspectives. Within this context, what Susman and Dean call "codification of information" is apt to be a more natural fit. On the other hand, if Susman and Dean are right, and successful DFM requires a combination of technical codification of information plus group processes and integrative mechanisms, traditional mechanistic organizations will probably need to make supportive changes in their structure, political context, and culture.

The main observation we have made is that when attempting to change the design of products so they are better integrated with manufacturing system design, change is needed on multiple fronts. Specific technical tools and organizational fixes such as design standards and cross-functional teams cannot simply be plugged into the organization and expected to make significant changes in the design process. This has been observed by others, including Adler (1989), who found that technical integration of CAD/CAM must be accompanied by appropriate social integration mechanisms, and Lawler (cf. p. 605, 1988), who argues that effective participative management depends on simultaneous changes in reward systems, information systems, authority patterns, and worker skills. Real people in real organizations are influenced in their goals and priorities by reporting lines, departmentation, what they perceive as their own interests, and the underlying assumptions they learn through social interaction. We have observed in the two case studies a number of organizational context barriers that inhibit practices that are considered desirable for DFM, e.g., goal consensus, patterns of rotation between design and manufacturing, and project continuity. There were also indications that intentional actions taken by management of these two divisions were succeeding in changing the organizational context to one more supportive of DFM. We briefly discuss what we learned within each of the three realms of formal organization, politics, and culture below.

Formal Organization and DFM

It has long been known in the organizational design literature that extensive differentiation must be accompanied by appropriate integration for organizations to be effective (Lawrence and Lorsch, 1967; Galbraith, 1977). A common approach has become the use of cross-functional teams. The composition of cross-functional teams is a complex issue, and the "right people" for a team will vary from organization to organization. We have seen that "product engineers" are often not the main designers of products. In many companies product or component engineers act as coordinators, but the core design work is done by "designers" and drafters. Thus, all three functions should be represented on the team. In fact, the division of labor is generally much more extensive than that between a product engineer and process engineer, and additional team members that may be needed include engineering analysis specialists, manufacturing process engineers, assembly engineers, supplier and vendor representatives, finance, purchasing, and marketing. Generally, designs of component systems proceed separately. At AUTOCO, separate cross-functional teams were used for each engine component down to as fine a detail as a design team for connecting rods. Thus, many teams are operating in parallel and need to coordinate their efforts across interconnected component systems.

Cross-functional teams can easily become a quick fix for an organization that does not want to change in any fundamental way. Lawler (1988) refers to parallel participative structures as structures, such as quality circles, which take place outside of the usual day-to-day structure of the organization. In this sense cross-functional teams are a parallel structure. It is easy to hold meetings across functions, but much harder to change the basic structure of the organization. Both Engine and Auto Parts made significant changes in structure, as shown in Table 13.2. The interviews suggested that these were crucial in providing a context within which innovations like cross-functional teams could function.

Although the formal structure had changed a great deal at Auto Parts, the actual practice of DFM lagged far behind that of Engine. In fact, by moving product engineers to the manufacturing site at Auto Parts, they seemed to be getting increasingly pulled into the reactive, firefighting mode of operation of the plant. That is, rather than a new organizational form emerging that emphasized product and process innovation, the product engineering organization was being absorbed into the traditional manufacturing bureaucracy.

A large part of the problem Auto Parts was having in mounting a serious development effort was due to difficulties in getting the resources they needed. DFM requires additional resources early in the design process, which in theory should be recouped later in reduced product costs, warranty costs, and the like. Unfortunately, these benefits are not easy to quantify on a project-by-project basis. As a new, highly visible product development program, Engine was able to get the resources they needed. This was not the case in the more routine Auto Parts operation.

Organizational Politics and DFM

Product designers are mainly concerned with product functioning. Manufacturing personnel are mainly concerned with ease of manufacturing and manufacturing costs. While working together on cross-functional teams seems to help each party appreciate the perspective of the other, we have seen little evidence that their goals become one and the same. Goal congruence can be facilitated by assigning product and process engineers to the same department. However, they still have different roles in the system and get rewards for success within their own functional domain. For example, to a product engineer excellent product functioning is intrinsically rewarding, even if his boss emphasizes manufacturability as a goal.

One interesting question is whether the elimination of differences in functional perspectives is desirable. Much has been written about conflict, and one important conclusion is that conflict can be a source of group energy if it is properly managed (Fisher and Ury, 1981; Schein, 1988). There is some support for a U-shaped theory of conflict that either too much or too little conflict is undesirable (Robbins, 1974).

In their classic study in organization design, Lawrence and Lorsch (1967) found that the most effective organizations combined differentiation with integration. Thus, eliminating differences between product and process engineers may not be the solution.

One supervisor at Auto Parts felt that their goal of creating a combined product-process design role was going too far. She claimed that you cannot create a "super engineer." Some people will be better at one side or the other, and there is strength in their different competencies. She personally had been a manufacturing engineer for two and a half years and a product engineer for ten years and felt her strength was overwhelmingly in product engineering. She felt she simply could not become a seasoned manufacturing engineer in the short time she spent in that function. There was also already some indication at Auto Parts of a conflict between the short-term pressures of current manufacturing and the need for a long-term focus on new product development. The experience of Auto Parts suggests that the short-term pressures will win out.

At another level of politics we found a conflict between the career goals of engineers and the requirements of DFM initiatives. We found component engineers struggling with the contradiction between what they knew was best for the design of the product and their own personal goals of climbing the corporate ladder. The system continued to reward generalist knowledge even after AUTOCO began to emphasize DFM. These cases suggest that for effective DFM, pay and promotion systems should provide rewards for engineers who wish to specialize in the product and process design of a particular component system comparable to what is available to those who take a generalist path to management. One approach would be to establish a kind of "dual-ladder" system, while avoiding the common pitfalls of this system (Wolff, 1987; Sacco and Knopka, 1983).

Much of the knowledge about part and equipment design is located outside of large manufacturing firms, which depend on a large number of parts suppliers and equipment vendors. DFM requires intimate knowledge of manufacturing processes— knowledge often possessed only by outside suppliers and vendors. The case studies suggested that DFM will be most effective if mechanisms are developed to reward outside parts suppliers and equipment vendors and knowledgeable engineers from these outside firms to be part of the design team from the beginning.

Organizational Culture and DFM

Both Engine and Auto Parts used co-location, which was described by several participants as one of the most powerful ways to break down cultural barriers. At Engine, the product organization historically was located only one mile from process engineering. However, as Tom Allen (1984) has shown us, when it comes to informal communication, one mile might as well be one thousand miles.

The different functions that make up cross-functional teams can be thought of as representing different subcultures with different languages, symbols, and values. Thus, to work together effectively as a team it is necessary to take time out from working directly on design activities to become a team. This includes learning each other's language and working through goal differences. Our case studies suggested that DFM will be most effective if cross-functional teams spend time educating each other about language differences and openly discussing goal conflicts. Perhaps conscious team-building exercises could help to uncover differences in functional perspectives and help members constructively use these differences to encourage creativity. It was also interesting to note that in this highly technical context a common technical language such as geometric tolerancing can do much to bridge the communication gap across functions.

Another aspect of culture is the social status we ascribe to the occupants of social roles. Product engineers have historically been thought of as "real engineers," and they have the credentials to prove it—bona fide college degrees in engineering. In fact, in our interviews respondents often slipped into talking about "engineering" as synonymous with product engineering. By contrast, manufacturing engineers come from the trenches of manufacturing. This was reinforced at AUTOCO by the personnel system, which ranked product engineers one grade level higher than equivalent process engineers. To encourage DFM, in the two sites studied process engineers were raised one grade level.

In this chapter we saw that there are many features of traditional bureaucratic organizations that create serious barriers to DFM, particularly within the context of a changing environment that demands some degree of product innovation. We are certainly not the first to notice that in a dynamic context more flexible, adaptive organizational structures and cultures are required. Thus, we suggest that such a change in the organizational context is a requirement for effective DFM. We have discussed some of the specific types of organizational context changes made in two divisions of a single company and how they influenced the success of their DFM programs. More rigorous studies are needed to understand better the ways in which organizational context constrains and supports DFM and the conditions under which different approaches to DFM work best.

NOTES

1. It may be useful to view organizational context as what is left over after we define a specific set of integrative mechanisms. For example, Susman and Dean did not define a supportive organizational culture that values teamwork as an integrative mechanism. We treat this as part of the organizational context. Yet if management were to attempt to manipulate the corporate teamwork values to encourage DFM, it would then be viewed as an integrative mechanism.

2. This account was provided by Janice Verkerke, a graduate student who had recently worked in product development at AUTOCO.

3. We have observed this and similar game-playing in numerous manufacturing plants, both in and out of the auto industry.

4. This is by no means ancient history in the auto industry. A recent article in *Metalworking News* (Rigley, 1990) discusses how a vendor lost a sole-source agreement with Chrysler.

5. We have seen a very similar division of labor at most of the other large companies that we have studied (Fleischer and Liker, 1990; Liker and Hancock, 1986; Liker, 1988).

REFERENCES

Abernathy, W. J. *The Productivity Dilemma: Roadblock to Innovation in the Automobile Industry.* Baltimore: Johns Hopkins University Press, 1978.

Abernathy, W. J., and J. Utterback. "Patterns of Industrial Innovation." *Technology Review,* June-July 1978, 40–47.

Abernathy, W. J., K. B. Clark, and K. Kantrow,

Industrial Renaissance. New York: Basic Books, 1983.

Adler, P. S. "CAD/CAM: Managerial Challenges and Research Issues." *IEEE Transactions on Engineering Management* 36(3), 1989, 202–216.

Allen, T. *Managing the Flow of Technology.* Cambridge, Mass.: MIT Press, 1984.

Allison, G. T. *Essence of Decision.* Boston: Little, Brown and Co., 1971.

Argyris, C., and D. Schon. *Organizational Learning.* Reading Mass.: Addison-Wesley, 1978.

Badawy, M. *Developing Managerial Skills in Engineers and Scientists.* New York: Van Nostrand Reinhold, 1982.

Bellah, R.N., R. Madsen, W. Sullivan, A. Swidler, and S. Tipton, *Habits of the Heart.* New York: Harper & Row, 1985.

Burns, T., and G. Stalker. *The Management of Innovation.* London: Tavistock, 1961.

Clark, K. B., and T. Fujimoto. "Lead Time in Automobile Product Development: Explaining the Japanese Advantage." *Journal of Engineering and Technology Management* 6, 1989a, 25–58.

Clark, K. B., and T. Fujimoto. "Overlapping Problem Solving in Product Development." In *Managing International Manufacturing,* ed. K. Ferdows. Amsterdam: North-Holland, 1989b.

Deming, W. E. *Out of the Crisis.* Cambridge, Mass.: MIT Center for Advanced Engineering Study, 1982.

Emerson, R. M. "Power-Dependence Relations," *American Sociological Review* 27, 1962, 31–40.

Fisher, R., and W. Ury. *Getting to Yes: Negotiating Agreement Without Giving In.* New York: Penguin Books, 1981.

Fleischer, M., and J. K. Liker. *Designers and Their Organizations: Social System Impacts on Design Quality.* Unpublished manuscript, Industrial Technology Institute, Ann Arbor, 1990.

Galbraith, J. R. *Organizational Design.* Reading, Mass.: Addison-Wesley, 1977.

Harbour, J. "Manufacturing 101 Revisited." *Automotive Industries* February 1987, 23.

Ishikawa, K. *What is Total Quality Control? The Japanese Way.* Englewood Cliffs, N.J.: Prentice-Hall, 1985.

Juran, J. M. *Leadership for Quality: An Executive Handbook.* New York: The Free Press, 1989.

Kamath, R. R., and J. K. Liker. "Supplier Dependence and Innovation: A Contingency Model of Suppliers' Innovation Activities." *Journal of Engineering and Technology Management* 7, 1990, 111–127.

Katz, D., and R. Kahn. *The Social Psychology of Organizations.* New York: Wiley & Sons, 1966.

Lawler, E. E. *High Involvement Management.* San Francisco: Jossey-Bass, 1988.

Lawrence, P., and J. Lorsch. *Organization and Environment.* Cambridge, Mass.: Harvard University Press, 1967.

Liker, J. K. "Survey-guided Change in Ship Design and Production: Prospects and Limitations, *Journal of Ship Production Research* 4(2), 1988, 81–93.

Liker, J. K., and W. Hancock. "Organizational System Barriers to Engineering Effectiveness." *IEEE Transactions on Engineering Management* EM-33(2), 1986, 82–91.

Liker, J. K., and R. J. Kamath. "Manufacturer-Supplier Relations and Product Design in U.S. and Japan Auto Industries." In *Handbook of Technology Management,* ed. D. Kocaoglu, New York: John Wiley & Sons, 1991.

March, J. G., and H. A. Simon. *Organizations.* New York: John Wiley & Sons, 1958.

Merton, R. K. *Social Theory and Social Structure.* New York: The Free Press, 1968.

Meyer, J. W., and W. R. Scott. *Organizational Environments: Ritual and Rationality.* Beverly Hills, Calif.: Sage Publications, 1983.

Mintzberg, H. *Power in and Around Organizations.* Englewood Cliffs, N.J.: Prentice-Hall, 1983.

Morgan, G. *Images of Organization.* Newbury Park, Calif.: Sage Publications, 1986.

Nadler, D., and M. Tushman. "A Congruence Model for Diagnosing Organizational Behavior." In *Organizational Psychology,* eds. I. Rubin and J. McIntyre. Englewood Cliffs, N.J.: Prentice-Hall, 1980.

Pfeffer, J. *Power in Organizations.* Marshfield, Mass.: Pitman, 1981.

Pfeffer, J., and G. R. Salancik. *The External Control of Organizations.* New York: Harper & Row, 1978.

Rigley, L. "Chrysler Puts Second Lamb Job Up for Rebid." *Metalworking News* 17(792), July 9, 1990, 1, 37.

Robbins, S. P. *Managing Organizational Conflict.* Englewood Cliffs, N.J.: Prentice-Hall, 1974.

Sacco, G., and W. Knopka. "Restructuring the Dual Ladder at Goodyear." *Research Management,* July-August 1983, 36–41.

Schonberger, R. J. *Building a Chain of Customers: Linking Business Functions to Create the World-class Company.* New York: Free Press, 1990.

Schein, E. H. "Coming to an Awareness of Organizational Culture." *Sloan Management Review* 25(2), Winter 1984, 3–16.

———. *Process Consultation,* 2nd edition. Reading, Mass.: Addison-Wesley, 1988.

Selznick, P. *TVA and the Grass Roots.* Berkeley: University of California Press, 1949.

Takeuchi, H., and I. Nonaka. "The New New Product Development Game." *Harvard Business Review,* January-February 1986, 137–146.

Weber, Max. "Bureaucracy." In *From Max Weber,* eds. H. H. Gerth and C. W. Mills. New York: Oxford University Press, 1946.

Wolff, M. F. "Revisiting the Dual Ladder at General Mills." *Research Management,* May-June 1987, 8–12.

14
THE ORGANIZATION AND MANAGEMENT OF ENGINEERING DESIGN IN THE UNITED KINGDOM

ARTHUR FRANCIS and DIANA WINSTANLEY

It is demonstrated elsewhere in this volume that speedy and efficient new product development seems associated with simultaneous engineering, multifunctional engineers, and project management. A great deal of the impetus for the adoption of these practices in the West has come from our developing understanding of Japanese practice. The MIT study of the world automobile industry in particular has done much to advance our understanding of these organizational and managerial strategies (Womack, Jones, and Roos, 1990).

In this chapter we report on a recent survey of the management of engineering design in a number of British firms. We discovered very few firms that were even attempting to adopt all three of these practices. Most of the firms we visited felt there to be deficiencies in the way they did new product development and managed the design-manufacture relationship, but few were seeking to apply organizational solutions. The most common response was to try to increase the skill levels employed in design by recruiting a higher proportion of graduate-level engineers. Those few companies that were trying to adopt project management, flexible use of engineers, and/ or simultaneous engineering were having some difficulties.

In this chapter we explore some of the reasons for these difficulties and argue that this trinity of organizational strategies (simultaneous engineering, multifunctional engineers, and project management) is, in Japan, embedded within a broader range of organizational features that innovative firms in the West need to take account of and respond to. We also examine in a limited way whether these responses are the only or best set of organizational/human resource management strategies for Western (or at least U.K.) companies and identify some alternative responses in both Germany and the United Kingdom. We suggest there may be a range of feasible alternatives influenced by the national and/or industrial context within which design for manufacturability (DFM) is carried out.

THE TRADITIONAL PATTERN OF ORGANIZATION AND MANAGEMENT IN THE U.K. ENGINEERING INDUSTRY

The empirical observations reported in this chapter were collected over a period of about 12 months in 1985–1986 during a series of visits to 30 U.K.-based companies in the engineering industry (loosely defined as employers of graduates from engineering schools). In most visits our data were collected primarily from an interview with someone in a senior position in the design function, usually either the director or manager. In a few cases this interview was followed up by a more intensive case study exercise, which involved interviews with a range of engineers in different functions. Other aspects of this work have been presented elsewhere, and we have observed in a number of our previous reports of this research (see, for example, Francis and Winstanley, 1989) that a craft, or artisan, form of organization seems to have persisted in British drawing offices (and in the engineering industry more generally) until very recently. It is only in the last two decades that this pattern has been superseded, and even then not totally. In this chapter we trace some of the reasons for this "modernization" of work organization, attempt to identify the models of work organization being used by those implementing change, examine the appropriateness of these models for the U.K. context, and suggest some possible ways to proceed. One dominant model of "modern" management and organizational practice is that which a number of British managers believe to be Japanese practice. We look in some detail at the components of this model and some U.K. firms' attempts to adopt this model.

The traditional model that has characterized British engineering management contains a number of interrelated features. Although there is some level of agreement over the nature of these, there is still some dispute over their causes and evolution. Its central feature is its craft/artisan nature, demonstrated both by the training and qualifications of the skilled staff and by a strongly functional form of organization structure. We summarize some of the important differences between this traditional model and what we have called the "modern" model in Table 14.1.

Long after other industrialized countries had created an infrastructure for their engineering industries that put at the top such institutions as the *grandes ecoles* (in France), the *technische hochschules* (in Germany), or leading engineering schools such as MIT (USA), the British engineering industry was relying for the bulk and core of its expertise on skilled craftsmen on the shop floor and craft-trained technical staff in the drawing office—the latter until at least up to 1939 being recruited from among the best of the shop-floor workers. This craft basis had significant implications for the level of expertise available within the company. It was the artisan's rule of thumb and reliance on past practice rather than the analytical skills of the university graduate and his/her access to scientific data that was the knowledge base in many companies. It is even doubtful that graduates used many of the skills and knowledge gained from their university education (Crawford, 1989). What is probably at least as important overall, and more important with respect to our interest in forms of organization, is that craft-based systems of production are usually characterized by high levels of occupational control over the work process. This has been extensively documented on the production side of the engineering industry (Friedman, 1977, for

Table 14.1. Stereotypes Compared of Traditional and Modern Design for Manufacture

Traditional	Modern
Organizational Structure	
Occupationally based	Managerially controlled
Functional	Project-driven matrix
Design-Manufacture Process	
Sequential/iterative	Simultaneous/parallel
Skills and Labor Market	
Fragmented	Integrated
High demarcation	Low demarcation
Product Market	
Slow moving	Fast moving
Expensive customization for long-term customers	Quick-response customization to changing range of customers
Educational Background of Designers	
Apprentices	Graduates
Culture	
Craft	Managerial/professional
Design Within the Organization	
Design conducted within large bureaucratized firm	Extended network of subcontracting relationships
A declining population of mature middle sized craft based engineering firms	

example) and our impression is that this has been similarly true on the design side. This high level of occupational control meant that managerial arrangements for controlling the work process were correspondingly weak and that the organization structure was grouped strongly around specific functions within the enterprise as occupations held sovereignty over their own functions with relatively weak interfunctional links.

This has been changing, but it is only in the last couple of decades that craft-trained technical staff have lost their predominant position to graduates. This has been documented by authors such as Ahlstrom (1982), Smith (1987), Watson (1975), Whalley (1982) and others; and there is evidence of it in the statistics on the educational composition of the engineering industry workforce. For example, the U.K. Engineering Industry Training Board's (EITB) annual figures (shown in Table 14.2) show a drop in technicians in the engineering industry of 21% from 1978 to 1990, and yet over the same period there was an increase of 64% in the number of professional engineers and scientists (all of whom would be graduates or equivalent). Such figures are reflected in figures presented in the Finniston Report (1980), which showed for younger compared to older engineers a move away from technician modes of qualification (Higher National Certificates and Diplomas) toward degree-level qualifications. The change also is evidenced by firms changing traditional job titles, which signified an occupational-type form of organization, such as chief draftsman, and chief engineer, to more managerial-sounding titles, such as design manager, engineering director, and so on. A number of quite senior engineers lamented this change.

We already have noted the strongly functional form of organizational structures. There are often rather many different functional groups in any one company, each quite narrowly specialized. It is common to find separate research, development, design, test, production engineering, production, and quality assurance departments, each rather autonomous. Engineering design itself is usually very narrowly defined as largely a drawing office activity conducted by draftsmen under the supervision of a chief draftsman. As one engineering manager said, "Design is everything associated with the production of drawings." The definition of other functional activities can be even more varied, particularly "development," which to some companies represents an activity associated with research at the front end of the design process, and to others is more of a "test" operation playing a larger role after the design input.

It is partly the craft and function-based form of organization that underpins the now much-criticized sequential design process. We and many others (see Winstanley and Francis, 1988) have drawn an analogy between this and a relay race with a large wall existing between the design and production engineering functions over which the design is passed. In reality in the U.K. the relationship between design and manufacture has traditionally exhibited an iterative process, more akin to a tennis match, where the design is passed to and fro between the respective departments.

The evidence from our survey is that this traditional model of organization and management in the engineering industry is now breaking down. It is not yet clear how it is being replaced. Most of the companies we visited were at an early stage of making changes. We chart here what we take to be the main factors leading to the breakdown of this traditional model and turn later to both the difficulties of replacing it with the dominant Japanese model and the question of whether this Japanese model is the one best way.

There seem to have been major changes over perhaps the last two decades. Some of these have been fairly long-term, rooted in the post-war development of U.K. society. Others have been more recent, triggered by changes in the international economic environment over the last 10 to 15 years.

We would highlight three of these new recent major changes. One is the increased awareness by U.K. firms of the force of international competition. U.K. manufacturing output has been in decline, as a proportion of total output, over the long term, as has its productivity relative to international competition. But superimposed on this long-term decline was a major economic crisis in Britain between 1979 and 1981 which caused a very sharp fall in output and profitability of the U.K. manufacturing sector. This resulted in some notable British engineering firms' making their first ever trading losses and exacerbated the decline of parts of the U.K. car industry. Though academics and policy analysts had been drawing attention for some time to the weak competitive state of much of British industry, it took the events of 1979–81 to send the shock waves throughout industry that triggered the awareness of the need for change and increased competitiveness that we observed in our study. Additionally, many U.K. firms became more exposed to international competitive

Table 14.2. Occupational Change in Engineering Industry

Occupational category	1978	1990	Change
Professional engineers, scientists, and technologists	58,256	95,697	+64%
Technicians and technician-engineers, including draftsmen	212,230	167,654	−21%
All staff	2,938,833	1,871,970	−36%

Source: EITB Statutory Returns.

pressure because of government policy changes about procurement and regulation. Previously cozy relationships between U.K. suppliers and industrial customers were broken up as companies either chose, or were forced, to source internationally rather than locally. This put U.K. engineering suppliers in particular under further competitive pressure.

A second source of pressure has been technological change. It is only since the late 1970s that computer-aided drafting/design (CAD) and computer-aided manufacture (CAM)—i.e., computer numerically controlled (CNC) machine tools and robotics—have been taken up to a significant extent within U.K. firms. There also has been during this period, partly because of this new technology and partly because of the extra competitive pressure we have just noted, demand in the market for more sophisticated products. Companies told us that it was partly in response to these technological pressures that they had stepped up their recruitment of graduate-level engineers. Certainly the figures for the employment of graduate-level scientists and engineers in the engineering industry in the U.K. are striking, as is shown in Table 14.2. We already have noticed the increase of over 60% since 1978 in the number of graduate-level technical staff employed. The picture is even more dramatic if one takes account of the decline in the overall numbers employed in that industry over this period. As a proportion of total employed, graduate engineers and scientists have increased by over 200%.

A third impact on British business in the 1980s has been the increased awareness of and interest in what competing nations have been doing, particularly Japan. Ford Motor Company's well-documented ''After Japan'' program was one example of this. Other companies have been similarly influenced. At the same time American consultants and management gurus also have been promoting new management practices that have attracted widespread interest; Peters and Waterman's *In Search of Excellence* (1982) and what followed being perhaps the best known example. Most recently the MIT world automobile study has increased knowledge of Japanese new product development and manufacturing practices within the automobile industry itself, including component manufacturers, not just from publication of the findings but from feedback to the industry from the research team while the study was in progress.

Our own interpretation is that these factors—increased competition, technical change, and awareness of new management practices—were three of the more important drivers for change in the organization of engineering design and new product development within the companies we studied.

There are, of course, other longer-term trends and these too are of importance. One of these has been the expansion of higher education in Britain. This, it is widely believed within the U.K. engineering industry, has led to a decline in the quality of people applying for apprenticeships (which we would here define broadly to include both craft and technician training based on programs that combine both practical and academic inputs), as those who would have come into the engineering industry via this route 30 years ago now have a greater opportunity to enter higher education. More recently, the introduction of government schemes such as the YTS (Youth Training Scheme) and previously the YOP (Youth Opportunity Program) is believed by some to have undermined rather than expanded the quality and quantity of apprenticeship training.

It may be that the decline in numbers entering through the apprenticeship system in the United Kingdom is more damaging than the changes in the raw numbers of entrants would suggest. A common criticism of the higher education route is that the courses, particularly in the leading universities, are too science-based and theoretical, reflecting the peculiarities of the British class structure, the values of which downplay the status of practical knowledge. Apprentices, drawn mainly from working-class backgrounds, have been thought to have practical experience and lack theoretical knowledge, with the reverse being the case for graduates from mainly middle-class backgrounds. More recently there have been attempts to correct this through extending graduate courses to include periods of industrial placement in enhanced four-year engineering degrees, a course of action recommended by the Finniston Report (1980), which identified as too analytical the majority of university engineering courses at that time.

A further difficulty posed by the break-up of the traditional model and the rise in numerical importance of the graduate engineers is the low professional status of the latter in the United Kingdom. Indeed, this low professional status has always been an issue, even within the traditional model. Various explanations have been advanced for this low status. Watson (1975), for example, argues that one reason for the traditional lack of status of engineers in Britain is the proliferation of professional associations representing engineers. The profession "engineer" in England dates from the formation of the Society of Civil Engineers in 1771 and has developed and fragmented to embrace over 30 major engineering institutions and over 150 if regional and local associations are included. It also might be argued that this fragmentation is itself the result of a political battle lost by the engineering occupation for full professional status of the kind enjoyed in Britain by the law and medicine.

The concern with status may be a very British obsession, or may simply reflect the view that engineers are less valued, with regard to pay, status, and access to positions of power both within and outside the engineering industry, than other professions in the United Kingdom and with engineers in other countries. It is also linked to the nature of the British class system, the structure and development of higher education, and the bureaucratization of the engineering profession. As Kindleberger has noted in comparing Britain with France (though the point could be made equally well in comparing Britain with Germany), "The point of contrast is that scientific and technical education were approved for the elite in France, by the elite in Britain" (1978, p. 235).

There are also those who argue that there has been a process of task fragmentation going on within engineering industry. Smith (1987) argues that the craft tradition has been eroded and tasks fragmented as part of a more general process of deskilling by management in order to erode occupational control. He cited, as an example of this, data from the Technical, Administrative, and Supervisory Section (TASS) of the Amalgamated Engineering Union (AEU). Their list of "typical posts" held by the membership amounted to 468 in 1973 as opposed to 7 a decade earlier in 1963. How much this proliferation of job titles represents a deliberate managerial strategy of fragmentation as opposed to a more reactive response to growing technical complexity must be in some doubt, but it is clearly a move in a direction opposite to that of multi-skilling, or the creation of polyvalent workers. Moreover, this fragmentation of tasks and skills has continued the tradition of a highly demarcated occupational structure with little movement between the boundaries. Although currently some engineers do move from design to development, for example, it is less common for engineers to move from design to production or to marketing as part of an organized career development path, except in some of the larger companies that have more recently created management development schemes to guide the careers of graduate trainees.

There is now a development away from task fragmentation in the form of a call for more flexible working. This has been linked to a number of different ideological debates, by no means all of which are influenced by modern Japan practice. This issue and the merits of functional flexibility have been debated by Atkinson (1984) and Pollert (1988). Atkinson and Meager (1986) in a survey on the extent of flexible working in Britain found that 90% of employers in the manufacturing sector were seeking an expansion in this direction. Flexible working is being debated in other countries too. An interesting development is the linking of changes toward greater flexibility in the product market with these changes in the labor market. For example, Kern and Schumann (1981) argue that changes in the product market—with a greater demand for customization, higher quality, speed to the market, and so on—have created pressures toward greater flexibility in skills in the labor market as firms try to develop a more adaptive work force.

A final point with regard to the traditional model and its breakup, and one we shall return to later, is the role played by subcontracting within the British engineering industry in general and design practices in particular. Internal contracting was widespread in the late 19th century, with Smith (1987), for example, documenting the extent to which craft workers would either work on their own account within larger factories, renting space and access to the power-shafts for driving their belt-driven machinery, or would carry out work for the factory owner/entrepreneur on the basis of a contract negotiated for each individual piece of work. The craftsmen would employ their own small teams to carry out these contracts. It was not until the early 20th century that a more bureaucratized model emerged, with engineers becoming embedded in large-scale organizations. This process accelerated between the first and second world wars (Watson, 1975).

There is now some evidence of this process going into reverse (Francis and Berkeley, 1989). Subcontracting relationships have always existed in the design area, and in particular there are a number of independent drawing offices servicing local large firms, but although the model of design as an activity internal to the large firm still persists today, there is evidence of numbers of firms extending their networking relationships and buying-in designs and/or design expertise. Our own survey turned up examples of this. One of the companies we visited had recently required its design office to tender for work in competition with outside companies, and another had allowed its design department to tender for outside contracts. Many of the companies used external contractors either to do routine design to cope with peak loads or, increasingly, to do specialized work outside the competence of the company commissioning the work. There were other examples of research and development being conducted on a subcontract basis. For example, some of our companies commissioned research by educational establishments, by high-tech companies on newly created science parks, or in joint ventures between education and industry.

It is not yet clear to us whether this apparently growing externalization matches the variety of subcontracting relationships that exist in some other countries and which support an extended and profitable small and medium-sized firm sector. In Germany, for example, there are research institutes in symbiotic relationships with small and medium-sized firms in some localities (Mickler, personal communication). This mirrors Herrigel's finding (1989) reported below.

While the most progressive of the firms in our survey had begun to take serious account of the break-up of the traditional model and had attempted to implement more modern models of organizational design—and we look in more detail at their activities below—a number of our firms seemed unable to escape. We came across several small and medium-sized engineering firms that had existed in a market niche for many years but had more recently found that declining profits had resulted in a cycle of takeover activity, lack of investment, and further lack of competitiveness and profitability. One such company that had existed for many years in the traditional British mold made conveyor chains. After years of decline they had been the subject of a takeover. This resulted in redundancies, reducing their work force from 500 to 250. Their sales declined, and they needed to diversify into other equipment to survive but were finding this difficult owing to a lack of investment or skills. Their organizational structure and culture reflected an image of a decaying traditional engineering company, as ad hoc redundancies distorted the structure, young graduates left, and an aging population of staff who had come through the apprenticeship route remained. The premises were that of a Victorian brick-built factory in need of modernization. Since we visited the company, they have been taken over yet again.

Our impression is that this traditional model of engineering in Britain is not confined to the mechanical engineering sector, which formed the bulk of our investigation. Other research has found similar features elsewhere. For example, Swarbrick (1990) found a very similar pattern in a U.K.-located television receiver manufacturing company. In this particular case the basic electronic design was done on "modern" lines, but the production engineering had been done on craft lines until just before Swarbrick's research commenced. His dissertation describes some of the difficulties management experienced in trying to move from this craft basis. In brief, their response, not unlike some of the engineering companies we studied, was to attempt to increase their expertise by recruiting young graduates into the firm but employing them within the preexisting form of work organization. Certainly there are differences between this firm and some of the newer electronic and computer companies, but the maturer electrical companies that have moved into electronics have been grappling with similar problems and the need to change the structures, cultures, and skills of their organizations to improve design for manufacturability.

GENERIC PROBLEMS FOR U.K. FIRMS IN ADOPTING SIMULTANEOUS ENGINEERING, MULTIFUNCTIONAL SKILL STRUCTURES, AND PROJECT MANAGEMENT

The most sophisticated British responses to the changes we have noted above came from those companies that had gained some experience of Japanese management practices. We note below a variety of responses made by U.K. firms. In this section we look at some of the generic problems that we believe U.K. engineering managers face in trying to implement the three major elements commonly identified as the core of good Japanese management practice.

These three core beliefs were held by only a handful of British companies. The handful of firms in our survey that already espouse these three core beliefs were those that either directly faced fierce Japanese competition or had entered into joint ventures with Japanese companies. This in itself is a comment about the relative ineffectiveness of innovation transfer mechanisms.

These particular companies were aware, in some cases because of the Harvard/MIT work on new product development in the car industry, that Japanese firms did new product development more quickly and cheaply than British firms and that the first products on the market were of higher quality because Japanese companies had already sorted out the design before production started. The British firms were particularly conscious that they were still doing redesign to iron out problems even after production had begun.

Strikingly absent from the responses from our companies was mention of a number of other salient features of Japanese management practice, in particular the principle of continuous improvement (kaizen) evidenced by quality circles among other things, and the extensive use of subcontractors in design work.

Much more serious was the complete absence of any evidence that U.K. managers realized that these Japanese practices were embedded in a matrix of other organizational, managerial, and cultural practices and had to be understood in that context. We gained the impression that simultaneous engineering, multifunctional engineers, and project management, were each stand-alone organizational choices from, so to speak, a smorgasbord of managerial options.

Partly from the literature on Japanese management practice, and partly from a visit by one of us to Japan, which included interviews with management from three large engineering firms, we have come to the view that there is a series of important interconnections between the various features of Japanese organizational and managerial life. We have tried to chart some of these in Figure 14.1.

Figure 14.1 attempts to demonstrate that the organizational tactic of simultaneous engineering, which is handled by using "heavyweight" project managers to coordinate multifunctional engineers, is made possible because of the extent to which it fits a series of related organizational strategies.

Central to these is the organizational orientation of workers in the larger Japa-

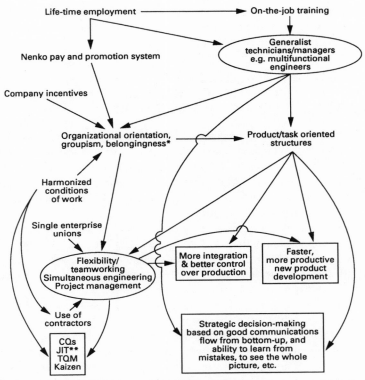

Figure 14.1. Some relationships between aspects of Japanese management practice.
*cf. Western occupation orientation or contractual relation with, and independence from employer.
**By JIT we mean not only just-in-time, but, following Voss (1986), the emphasis on maintaining flow, flexibility, and developing the chain of supply.

nese firms, with its associated group-ism and sense of belonging. This is to be contrasted with the more contractual orientation of many Western workers, including professional-technical staff, who value a sense of relative independence from their employers: an orientation, as we noted above, that in the past has often also included a strong element of occupational identity and control.

The Japanese orientation toward the organization is created and fostered, in part, by the lifetime employment pattern of the larger Japanese firms. This is buttressed by the *nenko* system, in which pay and promotion are awarded on a combination of seniority and merit, the former being substantially more important than the latter. To describe this as an internal labor market would do damage to the concept of the market. Decisions on the allocation of staff are based on very much more than internal price signals.

Lifetime employment and the *nenko* system mean that companies can plan individuals' career progression and training and thus provide the kind of general experience and extensive on-the-job training underlying what people seem to mean by the term *multifunctional* engineer. Generalist technicians and managers are said to be features of many Japanese companies. By this, commentators seem to mean that staff move in a planned way through all the main departments of the firm in the early years of their career so that they have a good working knowledge about every aspect of the firm's business. This is in contrast to the job-hopping pattern of moves between firms but within one particular functional specialty more characteristic of contractually and occupationally oriented Western high-flying technical and managerial specialists.

Because of the organizational orientation and generalist skills of Japanese staff, it may be easier for Japanese than Western companies to organize along product or task lines compared to functional lines. Because of a greater degree of shared knowledge about the product and task among Japanese staff, communication and coordination costs are lower and can therefore be handled laterally. There also is less commitment to one's functional/occupational identity and expertise and therefore less reluctance to be separated from a functional structure working with colleagues from the same functional background and for a functional boss who can protect one's functional interests. Moreover, if relationships between staff are known to be very long-term, one would expect to find less of the short-term opportunistic behavior corrosive of lateral relations in organizations.

The organizational orientation and the product/task-oriented structures enable greater flexibility of operations (less job demarcation) and more teamworking and all these enable the simultaneous engineering so well described by the Harvard team (Clark, Chew, and Fujimoto, 1987; Clark and Fujimoto, 1988; Clark, 1989).

Associated with these internal arrangements are the subcontracting relationships common between Japanese firms. As the Harvard studies show, the extent of subcontracting, even of design work, is markedly more extensive in Japan than the West. The evidence seems to be that such subcontracting is not of the arm's-length adversarial kind still common in British industry but that which Williamson (1975) would describe as relational: client Japanese companies typically employ one subcontractor over a number of years for any particular task. This seems to be an extension into the firm's environment of the organizational/groupism/belonging orientation found within the firm. Because of this, flexible, fast, and well-integrated teamwork can take place between firms and their contractors as well as within the client firm.

This difference between Japanese and Western organizational forms and managerial practices was graphically described to one of the authors when he visited the main Japanese factory of a firm with an important joint venture with a U.K. company. The visit was at a time when the U.K. company had just launched the U.K. version of the Japanese product. The Japanese version was manufactured in Japan and had been launched some months previously. The U.K. model was being manufactured in the U.K. and was to be made in two versions—one was labeled with the Japanese company's name, the other with the U.K. company's name. The Japanese company therefore had a strong interest in a successful product launch. There were, however, problems, and this was confirmed by staff from the U.K. company that the author met on his return to Britain.

These problems were relatively minor ones, requiring slight modification work to processes and parts to ensure high-quality assembly and finish and were to be expected in the run-up to launch of any complex product. What bewildered the Japanese was the difficulty the British seemed to have in getting the appropriate corrective work done. In Japan, they told us, if there was a problem relating to one part of the production process, the person in charge of that part of the factory would have within his or her staff someone who could deal with that problem. Departmental managers would have a fair working knowledge of what was going on because of their generalist skill resulting from the extensive on-the-job training received in their early years in the company. This means they would be able to carry out some sensible analysis and diagnosis, alert the appropriate staff member within the department, brief them sensibly about the issue, and exercise responsible supervision of their actions.

Our Japanese respondents believed that in the British case, by contrast, the person in charge of the department would be unlikely to have detailed knowledge about a significant number of the internal processes of that department, partly because his or her previous career might have been in different departments and, quite likely, outside that firm. It would be difficult therefore for the person to lead an analytical-diagnostic exercise to identify the cause of the problem. In addition, the expertise to deal with the problem was likely to be held not by one individual but to be shared by a group of different people, most of whom probably belonged to functional groups not under the control of the head of the department in which the problem arose. This resulted, we were told, in the U.K. company's never making a proper diagnosis of its problems and spending a great deal of time trying to organize a response and solution. So infuriated was our Japanese informant as he tried to chart these difficulties on the blackboard that he broke his chalk as he tried to write!

U.K. FIRMS' RESPONSE TO THE JAPANESE MODEL

Given the generic difficulties we have suggested facing U.K. firms attempting to implement elements of the Japanese model, how did the U.K. firms in our sample fare in their response to it?

First, where did their information come from? We found there to be three main sources of information:

1. from comparison with Japanese competitors or joint venture partners;
2. from the recruitment of senior managers with experience working for Japanese firms, or firms that have adopted some aspects of Japanese practice; and finally,
3. from the popular academic literature and management consultants (often via seminars).

For example, one firm, which we shall call Compo for anonymity, had tapped all of these sources (a case study based on Compo is written up by Winstanley and Francis in Wilson and Rosenfeld, 1990). Compo is a large, British, unionized multinational that produces electrical and mechanical components for a range of industries. It had for a long time exhibited many of the characteristics we have identified with the traditional British model. The first pressure on the company was the increasingly stiff competition in the marketplace, the U.K. side of which was contracting in size and subject to intensifying foreign competition. A corporate-level task force was set up to investigate Compo's competitive position. This produced a number of recovery achievement strategies where each part of the business was compared to the best competition in that area. The task force found that although Compo compared favorably on product quality, and had a strong emphasis on manufacturing competence, it was vulnerable in its manufacturing costs and product lead times.

The second force for change was the recruitment of two key influential managers, who became "change agents." One of these, who was put in charge of driving through changes in the design area, had worked for a company that had adopted some variants of the Japanese model. He believed "the British desire to rush into hardware is the Achilles' heel of British industry" and wanted to overturn Compo's traditional "cut and try" methodology. To do this he challenged each feature of the British stereotype, attempting to develop a more parallel approach to product and process design, seeking more teamwork at the front end of the design process, in order to reduce design iterations later on and costly changes to hardware. He wanted to develop broader designers and move away from design specialists toward design generalists, with the eventual aim being to create multifunctional engineers. For the organizational structure, he propounded moving from a functional-based matrix to a product-based matrix where the majority of staff would be physically placed in project teams.

A third change strategy was the collection by senior managers of many of the popular texts on new-style management and attendance at seminars given by management consultants. One of Compo's chief engineers in charge of implementing major changes in the business organization of the design function, collected all the "management by checklist" texts. An illuminating example of the way certain texts drawing on what is believed to be Japanese practice can become a design manager's bible is furnished by one memo that he sent round to all the engineering managers in the company, enclosing extracts of such a text, Pilditch's *Winning Ways*. It said:

> In summary, the speaker presented nothing that others, i.e., Peters and Waterman, have not already told us. He acknowledged this, but stressed that if companies actually implemented the things he described—then their performance would surely improve . . . . The attached summary of . . . [the book] contains most of what we are currently attempting—it reinforces the direction but [inevitably] does not say "how"—that remains up to us. . . ."

In implementing the organizational and management changes Compo came up against a number of problems. These related as much to the method of implementation as to the direction of the change. Technical specialists feared loss of expertise, and career progression routes became less clear. Not surprisingly there was opposition from functional heads, who perceived a loss of status and role. There was a difficulty in getting business units to accept a top-down change imposed from above, and engineers distanced themselves from the change due to lack of consultation. These were exacerbated by credibility problems, as the organization had a history of failed attempts at organizational change, and a "low trust" culture. The move toward the Japanese model was further thwarted by the promotion of the senior manager—the architect for the changes in the design—to another position and the firing of one of the engineering managers responsible for implementation.

At other companies there were also some fears expressed over designers' losing their specialist expertise in this drive toward functional flexibility. One manager pointed out that the Japanese development system did not advocate total multifunctionality among engineers, many of whom would work in very specialized areas; this would be obtained only through long-term career development in the company, where an engineer would be rotated through a number of specialist jobs, resulting in an expansion rather than dilution of specialist skills.

Not all companies had the same model of the future. Another company involved in manufacturing food-processing equipment and also trying to modernize had a different approach to broadening the skills of design engineers. One engineering manager from this company at a seminar we organized to disseminate some of our research findings engaged in heated conversation with the manager from Compo over his model. He argued that instead of trying to make designers broader, multidisciplinary, and multifunctional, companies should retain their specialist skills but develop multifunctionality through placing them in project teams with specialists from other areas such as marketing, production engineering, and so on.

Another nontraditional design consultancy company was not even prepared to move this far toward project teams. Here managers argued strongly for retention of a specialized functional departmental organization structure that would take precedence over a project team approach to design. In this case they believed that the developments in technology and technical knowledge required narrow specialists, and they were decreasing multifunctionality by putting their engineers into increasingly specialized roles. This firm believed it to be important to keep such engineers within their discipline-based departments because their clients were buying design expertise to develop prototypes. Therefore communication within specialist areas was more important than between different functions. The interface between design and manufacture became not just one wall, but several walls within each area. An additional communication interface was between designers within the consulting company and both the design and manufacturing engineers within the client company. In reality many of these walls and barriers were found to be nonexistent, and they believed this was because of the small size of the company, which had approximately 450 employees—200 professional engineers (mainly mechanical), 100 craftspeople, 80 technicians, and 70 support staff.

This company, which we shall label Bucardi for convenience, is one of a number of such companies that is beginning as part of the move by larger U.K. firms to subcontract activities to consultants and others, thus leading to a greater networking of relationships between firms. Such small specialist consultancies suffer from an undiversified range of activities, so that when the inevitable troughs occur within their industry's market they take the brunt of it. However, this company had still managed to retain a fairly "high trust" relationship with its work force, treating staff as professionals and, with 10% to 15% of their sales invested in R&D, they managed to provide financial support for creativity and innovation among their staff, backing promising ideas vetted by a technical committee.

SOME COMMENTS ON THE BRITISH ATTEMPTS TO ADOPT ASPECTS OF THE JAPANESE MODEL

At the very least, Western firms need to be doing more than simply bolting on a few simple organizational mechanisms. Appointing project managers, however senior and "heavyweight"; creating project teams; and instructing staff to communicate between departments earlier in the design process will not be enough to create organizations equivalent in speed and effectiveness to those found in Japan. The change process has to go deeper, to affect individual orientations to the firm and to the task, and few firms are attempting to put in the level of effort required to do this. It is not a short-term solution and is not cheap. Some of the major U.K. engineering companies are now investing a great deal more money in, for example, the training of design and production engineers at the postgraduate level. Hearsay evidence is, however, that at least some of the firms doing this are losing a substantial part of any benefit because they fail to retain the staff they are training. The effect of this enhanced training is to increase the attractiveness of those trainees in the external labor market, with obvious results. Also, unless firms change their patterns of work organization and career progression, enhancing the skills and expectations of individual engineers only results in their feeling underutilized. The impression one often has, and this is backed by a limited amount of data from our survey, is that some firms have attempted to respond to what they see to be a skills shortage in their firms by increasing the number of graduate engineers recruited, either absolutely or at the expense of technician recruitment, but then to do nothing about the kind of job the graduate does. Common practice seems to be to slot the graduate into the same job and structure previously designed for a technician-level engineer and then wonder why the graduate doesn't possess the same high level of practical skills as a technician, but also gets bored with the lack of intellectual challenge in the job.

Our conclusion is that for companies that want to follow the emulate-Japan route the tactical organizational and managerial changes must be accompanied by changes in the culture of the organization, softening structures generally and also overhauling the human resource management system. There must be coherence between the sources of recruitment, work organization, rewards, status differentials, and the training and career progression systems in use in the firm. This in turn requires far greater coordination and understanding between the engineers and the personnel function than usually obtains in engineering companies.

However, though it is clear that all U.K. companies must continue to strive to provide organizational structures, management techniques, and human resource management policies that match current requirements; and though it is equally clear that many of the firms in our survey were using systems and techniques that were in desperate need of updating, it is less clear that the Japanese model provides the one best way, suitable for all Western firms. At its heart the Japanese model appears to rely on a highly stable work force that can be centrally managed. In the language of political science, the large Japanese core companies that use the model described here are command economies and not market economies. Command economies are now out of fashion politically, and one of their difficulties is that they rely on a very high level of either assent from the citizenry or control from the state agencies. This seems to have been present in the large Japanese companies, at least since the 1960s, but it is less of a feature of Western firms. Western firms are, of course, attempting to generate these levels of assent and commitment through culture-change exercises, but there must be a question as to whether this is always going to be either the most effective, or even politically feasible, way forward. It is interesting to note, though we can go no further than that in this chapter, the possibility that changes in Japanese society as it becomes more affluent and "modern" are reducing the possibility of generating the kind of commitment by workers necessary for running a command economy within the firm. There is already a word in common use in Japan—*shinjin-rui*—that describes what may be considered a Westernizing phenomenon. The *shinjinrui* are literally the "new breed" of young highly qualified Japanese workers who are prepared to move from large company to large company in response to market signals, a willingness matched by an increasing desire by the large companies to recruit these early/mid-career specialists.

In the final part of this chapter, then, we take a brief look at some of the alternative possibilities that may be available to Western firms in organizing speedy and efficient new product development and design for manufacturability.

Some Lessons from Germany?

An interesting comparison with the Japanese ideal-typical case comes from work done on the German machine tool industry. Herrigel (1989) notes two alternative organizational forms extant in this industry. The one he describes as autarchic, the other decentralized-region-based. Neither corresponds to the Japanese model.

The autarchic form is self-contained, functionally organized and based to a considerable extent on occupationally controlled craft training with artisan control over the organization of work, the specialized workshop being at the core of the process. Or this was the organizational form until rather recently.

The decentralized-region-based form comprises a network of organizations contracting with each other in a relational way, but unlike the Japan case, the relations are undergirded by a strong moral order in the local residential community within which the firms operate.

Each operates within a different region within Germany, the former being more typical of the Ruhr Valley, Westfalen more generally, and in many of the old court and trading cities such as Hanover, Kassel, Nuremberg, Augsburg, and Munich, and the latter located in areas that have been dominated by the putting-out system or had been traditional centers of handicraft metalworking for centuries. In each case there seems to have been some kind of fit between the type of industrial order and the local educational arrangements.

Each form has been changing recently in response to some of the same kinds of pressures identified earlier in this chapter that are applicable to Britain—principally technical and market change requiring a shift to higher-quality flexible production. The autarchic form has moved to an extent from large organizations containing within them artisan-based workshops to smaller, specialized units. The decentralized-region-based form has expanded and diversified its network.

Much more information is needed to understand whether the former form has led to a breakdown of functional boundaries so that simultaneous engineering now takes place and whether the expanded and diversified network of the latter system can also handle this simultaneity, but the presence of this diversity serves as a warning against simplistic, catch-all organizational/managerial solutions.

The engineering industry in the United Kingdom is much more fragmented than it is in Germany. There is certainly not a machine-building industry of similar size. A broad impression would be that in the U.K. there are constellations of firms surrounding the automobile and aerospace industries, a certain presence in food-processing machinery, and various firms doing fairly basic metal-bashing, protected from overseas competition by the high ratio of transport cost to value-added. Beyond that there are pockets of industry with a presence in particular markets—for example, pumps and valves, and certain capital goods suppliers for basic industries such as mining and the railways.

The largest firms in the motor and aerospace industries resemble to a degree Herrigel's autarchic concept, though each group seems to engage in a high level of subcontracting. Half the added value of an aero-engine, for example, consists of bought-in parts, a significant number of which are "black box" items (i.e., with the core of the design done by the subcontractor). We are aware of the responses by these firms in terms of increasing the recruitment of graduates and attempting to "Japanize" some of their own management practices, but we have little knowledge on proactive measures at the industry level, except for a certain amount of consortium activity in post-experience training of the design and production engineers. Has there been, in Britain, in response to changing pressures to improve the design-manufacturing interface changes to the structure of the industry similar to those Herrigel observed in Germany? Are companies changing their networking behavior to create more relational contracting with suppliers and clients? What is being done at the industry level about education, training, and vocational qualification arrangements? What industry-level initiatives are being mounted to help create the necessary culture change in the industry? We have not directly addressed these issues in our research and so cannot speak authoritatively, but we suspect that there is a lack of proactivity on these topics. Industry-level changes may be taking place, partly reactively, and partly led by incoming Japanese companies.

Some Lessons from Recent Developments in Organization Theory?

Striking among the recent detailed studies of the design-manufacturing interface is the absence of any reference to organizational phenomena described in the post-1980s North American literature on organizational behavior. The various organizational arrangements described are those found in Galbraith's Organization Design (1977). Nowhere do there appear to be accounts of organizations adopting the kind of strong-culture soft-structure arrangements popularized by Tom Peters (1987) and Rosabeth Moss Kanter (1984, 1989).

In some ways this is surprising because, perhaps not coincidentally, the kind of organizational arrangements now being promoted by such as Peters and Kanter mirror rather accurately some key aspects of the Japanese system noted above. In particular, the emphasis on promoting a strong organizational culture is a Western parallel to the organizational orientation of members of Japanese companies. Firms that develop mission statements, talk of developing a sense of ownership of problems and processes, and get people to "sign up" for projects are clearly trying to shift individuals' orientation to the firm away from the arm's-length contractual, or occupational, relationship typical of the past and toward a sense of belongingness to the firm.

Similarly, softening of structures, when those structures have been functionally based and, probably, buttressed by a large element of occupational control, is a move toward product and task-oriented arrangements within which flexible, non-job-demarcated teamworking can take place.

These kind of processes are described, largely uncritically, in, for example, Tracy Kidder's account of the development of a new computer (1981) and, more extensively, in Kanter (1984, 1989).

The big question is the extent to which it is possible to generate a strong culture and soften up a functional structure simply by using process consultants within the enterprise, without changing structures such as the nature of the employment relationship (lifetime employment) and the extent of career planning and training within the firm. Much more important may be the strongly individualistic ethos of Western society and the various features of the society, such as the type of educational system, which lie behind this ethos and sustains and recreates it.

Another difficulty, particularly in the United States and the United Kingdom, for companies attempting to create high-commitment "command" style organizations are the uncertainties about employment and career prospects among staff in economies in which the manufacturing sector in general and the engineering industry in particular is in a weak position and highly vulnerable to swings in the general economic climate. If it is a common experience within the industry that firms engage in lay-offs at worst and thwarted career progression at best, then it will be difficult, as some of the firms in our survey found, to generate a high-trust, high-commitment culture. Nevertheless, the organizational issue for the 1990s and the next century will be how to coordinate the activities of highly qualified technical staff.

An urgent question both for academic researchers and for industry-based practitioners is how this should happen. It is not obvious that the way in which a late-industrialized country with agrarian origins (Japan) has been able to organize, based on a particular set of cultural and institutional arrangements that had their historical roots in a society dominated by a samurai class and a rice-growing peasantry, provides all the lessons for Western post-industrial society. This would be especially the case if it turned out that the Japanese automobile industry was one of the last mass-production industries, and one of the last to use a form of organization developed and refined in the 1960s–1980s and which by 1990 was beginning to be past its peak.

The idea that there may be some substance in these two fears—that the Japanese model does not entirely fit Western circumstances and that the Japanese model itself be at the point of decline—is given credence by recent developments in both academic organization theory and best-selling management texts. These developments reflect and build upon a great deal of work now going on within philosophy and social science under the label of post-modernism (see, for an excellent summary, Harvey, 1989).

To speak of post-modernism and engineering design may almost be a contradiction in terms. It is not possible to do justice to the idea of post-modernism within these closing paragraphs, but one could crudely summarize the central idea of post-modernism as follows: the modern hope has been, since the Enlightenment, that rational, purposeful, hierarchically organized activity is both possible and the most effective way of getting things done. We are now in a post-modern era, where we recognize many of the limits to this modern approach. Rationality is often, or even fundamentally, impossible, and the world is too complex and fast-moving to establish purposes and hierarchies. Things happen in all kinds of ways, often chaotically, sometimes creatively. The post-modernist view is both a critique of the modernist endeavor but also a prescription for a new way of doing things.

Design could be considered to be a classically modern, as compared to post-modern, activity; and yet current orthodoxy in organization design, at least according to Tom Peters and Charles Handy, is aggressively post-modern. Their two latest books have titles that are clearly influenced by the post-modernist movement. Peters's is called *Thriving On Chaos,* and Handy's, in unacknowledged reference to Kant, carries the title *The Age of Unreason.* Their emphasis is on openness, process, and participation rather than on closure, the finished product, and hierarchy. How then is design to be done in a post-modern world?

To the extent that it is helpful to describe the world as post-modern, what we might expect is an absence of a blueprint for the management and organization of successful design. There are certain elements of the Japanese model that have post-modern characteristics. The notion of *kaizen,* continuous improvement, is one. An engineer at one of the U.K. factories engaged in the joint venture with the Japanese described earlier spoke to us of a particular source of conflict between U.K. and Japanese engineers. His difficulty was locating the definitive design of any particular product. In the U.K. the definitive design was that set out in the most recent set of drawings. Any changes to the design had to be recorded on the drawings and that change approved. The approved drawing was the master. In the Japanese factory, by contrast, the definitive design was that which was incorporated in the most recent product coming off the line. The technical drawings were updated after the event and always trailed the latest design changes. This constant state of flux and change is characteristically post-modern.

However, what is also post-modern is the growing disaggregation of Western companies. The implication of post-modernism for the organization of design is that in the future design will not get done as a result of neat schemes of work organization backed by well-fitting vocational training schemes and tidy human resource management strategies. Good designs will emerge from a bricolage of all kinds of bits and pieces of expertise, experience, and networked knowledge. Facilitating rather than planning will be what is required, though know-how will be enormously important also. It is perhaps in this context that the findings both of the study reported here and that of Francis and Berkeley (1989) about the growth of subcontracting in the engineering industry should be seen. However, the choice is not just between a 1980s-style Japanese "command economy" organizational solution and a Thatcherite return to market arrangements. Our attention should now be focused on the management of networks between groups of activities, such networks having economic, social, and political dimensions.

REFERENCES

Ahlstrom, G. *Engineers and Industrial Growth.* London: Croom Helm, 1982.

Atkinson, J. "Flexibility, Uncertainty and Manpower Management." *IMS Report* No. 89, IMS, Sussex, England, 1984.

Atkinson, J., and N. Meager. "Is Flexibility Just a Flash in the Pan." *Personnel Management,* September 1986.

Clark, Kim B. "High Performance Product Development in the World Auto Industry." Harvard Business School, Working Paper 90–004, 1989.

Clark, Kim B., and T. Fujimoto. "Overlapping Problem Solving in Product Development." In *Managing International Manufacturing,* ed. K. Ferdows. Amsterdam: Elsevier, 1988.

Clark, Kim B., W. Bruce Chew, and T. Fujimoto. "Product Development in the World Auto Industry." *Brookings Papers on Economic Activity* 3, 1987, 729–771.

Crawford, Stephen. *Technical Workers in an Advanced Society: the Work, Careers and Politics of French Engineers.* Cambridge, England: Cambridge University Press, 1989.

Finniston M. "Engineering Our Future." Report of

the Committee of Inquiry into the Engineering Profession. London: HMSO, 1980.

Francis, A., and S. Berkeley. "Strategic Organizational Change for New Product Development in England." Unpublished paper presented to the *Ninth SMS Conference*, San Francisco, 1989.

Francis, A., and D. Winstanley. "Organizing Professional Work: The Case of Designers in the Engineering Industry in Britain." In *Competitiveness and the Management Process*, ed. Andrew M. Pettigrew. Oxford: Basil Blackwell Ltd., 1989.

Friedman, A. *Industry and Labour*. London: Macmillan, 1977.

Galbraith, J. *Organization Design*. Reading, Mass.: Addison-Wesley Publishing Co., 1977.

Handy, C. *The Age of Unreason*. London: Hutchinson, 1989.

Harvey, D. *The Condition of Postmodernity: An Enquiry into the Origins of Cultural Change*. Oxford: Basil Blackwell Ltd., 1989.

Herrigel, Gary B. "Industrial Order and the Politics of Industrial Change: Mechanical Engineering." In *Industry and Politics in West Germany*, ed. Peter J. Katzenstein. Ithaca, N.Y.: Cornell University Press, 1989.

Kanter, R. M. *The Change Masters: Corporate Entrepreneurs at Work*. London: George Allen and Unwin, 1984.

———. *When Giants Learn to Dance*. New York: Simon and Schuster, 1989.

Kern, H., and M. Schumann. "Limits of the Division of Labor, New Production and Employment Concepts in West German Industry." *Economic and Industrial Democracy* 8, 1981, 151–70.

Kidder, Tracey. *The Soul of a New Machine*. Boston: Little, Brown, 1981.

Kindleberger, Charles P. "Germany's Overtaking

of England 1806–1914." *Economic Response: Comparative Studies in Trade, Finance and Growth*. Cambridge, Mass.: Harvard University Press, 1978.

Peters, T. *Thriving on Chaos*. Basingstoke, Hampshire, England: Macmillan, 1987.

Peters, T., and Robert H. Waterman. *In Search of Excellence*. New York: Harper and Row, 1982.

Pollert, A. "Flexible Firm—Fixation or Fact." *Work Employment and Society* 2(3), 1988.

Smith, C. *Technical Workers, Class Labor and Trade Unionism*. Basingstoke, Hampshire, England: Macmillan, 1987.

Swarbrick, A. "Shortage and Utilization of Engineering Skills in the Electronics Industry: A Case Study in British Television Manufacturing." Ph.D. dissertation, University of London, 1990.

Voss, C. A. *Just-In-Time Manufacturing*. Kempston, Bedford, England: IFS Publications, 1986.

Watson, H. B. "Organizational Bases of Professional Status: A Comparative Study of the Engineering Profession." Unpublished Ph.D. thesis, University of London, 1975.

Whalley, P. *The Social Production of Technical Work*. Basingstoke, Hampshire, England: Macmillan, 1982.

Wilson, D., and R. Rosenfeld. *Managing Organizations, Texts, Readings and Cases*. Berkshire, England: McGraw-Hill, 1990.

Williamson, O. E. *Markets and Hierarchies: Analysis and Antitrust Implications*. New York: Free Press, 1975.

Winstanley, D., and A. Francis. "Fast Forward for Design Managers." *Engineering Designer* 228(3), March 1988, 133–135.

Womack, James, Daniel Jones, and Daniel Roos. *The Machine that Changed the World*. New York: Rawson Associates, 1990.

15
EPILOGUE

GERALD I. SUSMAN

This book suggests that companies that depend on superior product development capability for competitive advantage require an exceptional ability to organize, process, and learn from information that is related to product development. Most of the authors in this book argued or implied that the product development process is information intensive; that is, project personnel need to exchange a significant amount of information during the product development process. Several authors relied heavily on information-processing concepts in their analyses.

This final chapter provides a common perspective from which to view many of the ideas introduced in this book. This perspective suggests that many of these ideas relate to actions that contribute to an organization's ability to organize, process, and learn from information that is related to product development. These actions can contribute to the organization's product development capability by simplifying and clarifying information, by developing an organization's ability to process information, and by facilitating an organization's ability to learn. The following three sections cite illustrative actions discussed in earlier chapters of the book that can lead to improving an organization's product development capability. The authors who made specific reference to the actions cited are mentioned in parentheses.

SIMPLIFYING AND CLARIFYING INFORMATION

Information can be simplified by reducing the number of parts that make up a new product, by standardizing designs for use in related products, or by limiting the number of new technologies that can be introduced into a new product at any one time (Adler; Barkan; Clark, Chew and Fujimoto; Rosenthal and Tatikonda). Information also can be simplified by segmenting the design problem into modularized sub-problems, thereby reducing the complexity of each sub-problem and reducing interdependence between them (Shirley). Codification and computerization of data can simplify and clarify information that is used by project personnel (Susman and Dean), while the ability to codify and computerize such data depends, in part, on successfully completing some of the other actions cited above.

Information can be clarified or made less ambiguous by helping project personnel to agree on a common product concept and on project priorities and appropriate actions. Quality function deployment can help personnel achieve such agreement (Rosenthal and Tatikonda). Sufficient effort during the concept development phase can help project personnel start with a clear product definition (Ettlie) and a clear vision to guide the development of a product over successive product generations (Sanderson).

DEVELOPING AN ORGANIZATION'S ABILITY TO PROCESS INFORMATION

An organization can improve its ability to process information related to product development by organizing on a product rather than a functional basis. If it is not feasible for it to do so, then its information-processing capability can benefit from strong project managers. Reducing differentiation between functions or improving an organization's ability to manage differentiation also can improve an organization's information-processing capability (Slusher and Ebert; Susman and Dean). This can be accomplished by job rotation, co-location, and status parity between functions. The social, political, and cultural context of an organization can influence the degree of differentiation between functions and lead to resistance when trying to modify the conditions that encourage differentiation (Liker and Fleischer; Francis and Winstanley).

Bringing the right people together at the right time is a necessary condition for effective problem-solving, whether these people are vendors or customers (Liker and Fleischer), or manufacturing engineers involved during the appropriate product development phase (Coughlan). Another condition is use of coordination mechanisms, e.g., standardized rules, periodic or regular meetings, that are appropriate to the nature of the problem to be solved (Adler; Slusher and Ebert). Other conditions include training of personnel in appropriate DFM tools and practices (Rosenthal and Tatikonda; Sanderson) and in problem-solving and team-building skills. Training of this kind will not be effective, however, unless supplemented by changes in organizational structure and culture (Francis and Winstanley).

IMPROVING AN ORGANIZATION'S ABILITY TO LEARN

Many of the practices discussed in the previous two sections are facilitators of learning, especially learning that continually improves the fit between product and process and the speed with which product development personnel recognize mismatches between them and reduce or minimize these mismatches. This type of learning can be characterized as single-loop learning (Argyris and Schon, 1978). Another necessary ingredient is the motivation to improve performance continuously on these dimensions. Ever-increasing cost, quality, lead-time and performance standards can supply this motivation as can appropriate project-based rewards (Susman and Dean).

Single-loop learning is facilitated by maintaining continuity of personnel on the same project (Susman and Dean), or at least within the same division or plant (Liker and Fleischer), or by assigning the same personnel to successive generations of a product family (Coughlan). Learning is facilitated also by the ability to capture data

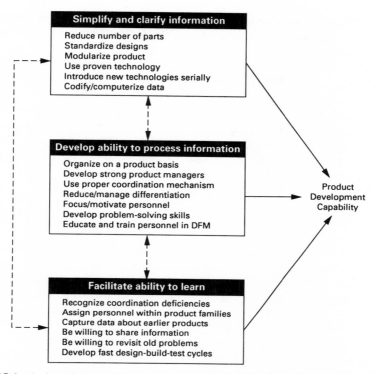

Figure 15.1. Actions that can improve an organization's product development capability.

about key attributes of each product in the family (Sanderson), by sharing common modules across members of a product family (Shirley), or by feedback on prototypes and dies from quick design-build-test cycles (Clark, Chew and Fujimoto). The willingness of project personnel to revisit old problems and share ideas about past successes and failures also facilitates this type of learning (Barkan).

Learning that leads to recognition that the coordination mechanism being used on a product development project is inappropriate for the problem being solved can be characterized as double-loop learning (Argyris and Schon, 1978). In this case, the conditions that facilitate problem-solving rather than the problem-solving process itself must be changed. These coordination mechanisms may have been inappropriate initially or technological or environmental pressures may have made them inappropriate. Product development teams will vary in the degree to which they can recognize that a deficiency exists and take action to correct it. Their ability to recognize a deficiency and act on it depends on the culture of their organization and on its available resources (Adler).

ACTIONS THAT CONTRIBUTE TO AN ORGANIZATION'S PRODUCT DEVELOPMENT CAPABILITY

Figure 15–1 summarizes some of the actions that contribute to an organization's product development capability. The top box includes actions that simplify and clarify the information to be processed. The middle box includes actions that develop an organization's ability to process information. The bottom box includes actions that facilitate an organization's ability to learn.

Actions in each of the three boxes contribute to an organization's product development capability. The arrows with solid lines that lead to product development capability indicate the direction of influence of these actions. The actions in each of these boxes may contribute additively to this outcome or the actions in one box may facilitate the effect of actions taken in another, i.e., their relationship may be multiplicative. The arrows with dashed lines between the three boxes indicate relationships between these respective boxes. Future research will determine the nature and direction of these relationships.

CONCLUSION

The theme of this book has been integrating design and manufacturing for competitive advantage. As the preceding chapters suggest, integration of design and manufacturing is critical to the success of strategies that seek competitive advantage through superior product development capability. Such integration allows design and manufacturing to process shared information efficiently and effectively so that high quality products can be developed more quickly and offered at competitive prices. The conditions that facilitate or inhibit such integration have been discussed as have the actions that the organization can take to improve its ability to process appropriate information. The success of these actions depends on understanding the role of information in the product development process and, of course, on the skill and vision of managers who are responsible for implementing these actions.

REFERENCE

Argyris, C. and Schon, D. A. *Organizational Learning*, Reading, Mass.: Addison-Wesley, 1978

NAME INDEX

SUBJECT INDEX